Plant Genetics and Genomics

Plant Genetics and Genomics

Edited by
Shane Simpson

Larsen & Keller
www.larsen-keller.com

Plant Genetics and Genomics
Edited by Shane Simpson
ISBN: 978-1-63549-223-1 (Hardback)

Larsen & Keller

Published by Larsen and Keller Education,
5 Penn Plaza,
19th Floor,
New York, NY 10001, USA

Cataloging-in-Publication Data

Plant genetics and genomics / edited by Shane Simpson.
     p. cm.
Includes bibliographical references and index.
ISBN 978-1-63549-223-1
     1. Plant genetics. 2. Plant genomes. 3. Genomics. I. Simpson, Shane.
QK981 .P53 2017
581.35--dc23

The publisher's policy is to use permanent paper from mills that operate a sustainable forestry policy. Furthermore, the publisher ensures that the text paper and cover boards used have met acceptable environmental accreditation standards.

Printed and bound in the United States of America.

For more information regarding Larsen and Keller Education and its products, please visit the publisher's website www.larsen-keller.com

Table of Contents

Preface

Plant genetics and genomics is a branch of biology which studies heredity and heredity transmission of characteristics in plants and other eukaryotic or multicellular organisms that are classified within the plant kingdom. This area includes various sub-fields like genetically engineered crops, genetically modified plants and examining the DNA and food safety and security. This book explores all the important aspects of this field in the present day scenario. It includes detailed explanation of the various concepts and applications of plant genetics and genomics. This book is compiled in such a manner, that it will provide in-depth knowledge about the theory and practice of plant genetics and genomics. Coherent flow of topics, student-friendly language and extensive use of examples make this textbook an invaluable source of knowledge.

A foreword of all Chapters of the book is provided below:

Chapter 1 - Plant genetics has many aspects. Plant genetics concerns itself with heredity and the characteristics inherited by plants. This is done through the study of the genome. The diversity among plants make this a vast field of study. This chapter will provide an integrated understanding of plant genetics; **Chapter 2 -** Plant genetics deals with the heredity of plants. The number of gene characteristics that are present in a species is known as genetic diversity. Topics included in this section include genetic diversity, genetic variability, plant tissue culture, biological hybrid and eugenics. The section strategically encompasses and incorporates the major key concepts of plant genetics, providing a complete understanding; **Chapter 3 -** The crops that are modified by using genetic engineering techniques are called genetically modified plants. It is mostly done to introduce certain traits in plants which are not naturally present in the plant. Some of the key concepts explained in this chapter are genetic variability, plant tissue culture, hybrid and eugenics. The chapter strategically encompasses and incorporates the major components and key concepts of plant genetics, providing a complete understanding; **Chapter 4 -** Plant breeding, like animal breeding, is introduced in plants to produce favorable characteristics in plants. A cultigen is a plant that is purposely selected by humans and happens because of artificial selection. The following text on plant breeding is an overview of the subject matter, incorporating all the major aspects of plant breeding; **Chapter 5 -** Chloroplasts are the specialized sub-units in plant cells and mutation is the perpetual modification of the sequence of genome of organisms or a permanent change in their DNAs. The other important features of plant biology are mitochondrion, polyploid and germline. Plant genomics is best understood in confluence with the major topics listed in the following chapter; **Chapter 6 -** The alteration caused in the genome of an organism by using biotechnology is known as genetic engineering. Genetic engineering techniques permit variation of the DNA of living organisms. There are many methods present in genetic engineering which have been listed in the chapter. It can best be comprehended in confluence with the major topics listed in the following section; **Chapter 7 -** Genetic erosion is a process where a threatened species of plant becomes extinct without a chance to breed with the same species. The term is used to describe the loss of specific genes or for even an entire species. Habitat fragmentation is another reason for genetic erosion. This chapter provides the reader with an in-depth understanding on genetic erosion;

Chapter 8 - Planting has been practiced by thousands of years. It can easily be traced to the beginning of human civilization. Modern scientific advances have enabled the study of plant genetics. This chapter helps the reader in understanding the evolution and history of plant breeding.

I would like to thank the entire editorial team who made sincere efforts for this book and my family who supported me in my efforts of working on this book. I take this opportunity to thank all those who have been a guiding force throughout my life.

Editor

Introduction to Plant Genetics

Plant genetics has many aspects. Plant genetics concerns itself with heredity and the characteristics inherited by plants. This is done through the study of the genome. The diversity among plants make this a vast field of study. This chapter will provide an integrated understanding of plant genetics.

An image of multiple chromosomes, making up a genome

Plant genetics is a very broad term. There are many facets of genetics in general, and of course there are many facets to plants. The definition of genetics is the branch of biology that deals with heredity, especially the mechanisms of hereditary transmission and the variation of inherited characteristics among similar or related organisms. And the definition of a plant is any of various photosynthetic, eukaryotic, multicellular organisms of the kingdom Plantae characteristically producing embryos, containing chloroplasts, having cell walls which contain cellulose, and lacking the power of locomotion. Although there has been a revolution in the biological sciences in the past twenty years, there is still a great deal that remains to be discovered. The completion of the sequencing of the genomes of rice and some agriculturally and scientifically important plants (for example Physcomitrella patens) has increased the possibilities of plant genetic research immeasurably.

Features of Plant Biology

Plant genetics is different from that of animals in a few ways. Like mitochondria, chloroplasts have their own DNA, complicating pedigrees somewhat. Like animals, plants have somatic mutations regularly, but these mutations can contribute to the germ line with ease, since flowers develop at the ends of branches composed of somatic cells. People have known of this for centuries, and mutant branches are called "sports". If the fruit on the sport is economically desirable, a new cultivar may be obtained.

Some plant species are capable of self-fertilization, and some are nearly exclusively self-fertilizers. This means that a plant can be both mother and father to its offspring, a rare occurrence in animals. Scientists and hobbyists attempting to make crosses between different plants must take special measures to prevent the plants from self-fertilizing.

Plants are generally more capable of surviving, and indeed flourishing, as polyploids. Polyploidy, the presence of extra sets of chromosomes, is not usually compatible with life in animals. In plants, polyploid individuals are created frequently by a variety of processes, and once created usually cannot cross back to the parental type. Polyploid individuals, if capable of self-fertilizing, can give rise to a new genetically distinct lineage, which can be the start of a new species. This is often called "instant speciation". Polyploids generally have larger fruit, an economically desirable trait, and many human food crops, including wheat, maize, potatoes, peanuts, strawberries and tobacco, are either accidentally or deliberately created polyploids.

Hybrids between plant species are easy to create by hand-pollination, and may be more successful on average than hybrids between animal species. Often tens of thousands of offspring from a single cross are raised and tested to obtain a single individual with desired characteristics. People create hybrids for economic and aesthetic reasons, especially with orchids.

DNA

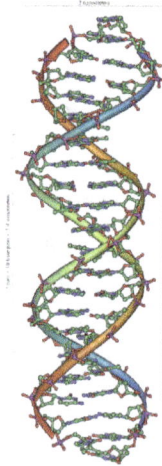

The structure of part of a DNA double helix

Deoxyribonucleic acid (DNA) is a nucleic acid that contains the genetic instructions used in the development and functioning of all known living organisms and some viruses. The main role of DNA molecules is the long-term storage of information. DNA is often compared to a set of blueprints or a recipe, or a code, since it contains the instructions needed to construct other components of cells, such as proteins and RNA molecules. The DNA segments that carry this genetic information are called genes, but other DNA sequences have structural purposes, or are involved in regulating the use of this genetic information. Geneticists, including plant geneticists, use this sequencing of

DNA to their advantage as they splice and delete certain genes and regions of the DNA molecule to produce a different or desired genotype and thus, also producing a different phenotype.

Gregor Mendel

Gregor Mendel was an Augustinian priest and scientist born on 20 July 1822 in Austria-Hungary and is well known for discovering genetics. He went to the Abbey of St. Thomas in Brno. He is often called the father of genetics for his study of the inheritance of certain traits in pea plants. Mendel showed that the inheritance of these traits follows particular laws, which were later named after him. The significance of Mendel's work was not recognized until the turn of the 20th century. Its rediscovery prompted the foundation of the discipline of genetics allows geneticists today to accurately predict the outcome of such crosses and in determining the phenotypical effects of the crosses. He died on 6 January 1884 from chronic nephritis.

Modern Ways to Genetically Modify Plants

There are two predominant procedures of transforming genes in organisms: the "Gene gun" method and the *Agrobacterium* method.

"Gene Gun" Method

The "Gene Gun" method is also referred to as "biolistics" (ballistics using biological components). This technique is used for in vivo (within a living organism) transformation and has been especially useful in transforming monocot species like corn and rice. This approach literally shoots genes into plant cells and plant cell chloroplasts. DNA is coated onto small particles of gold or tungsten approximately two micrometres in diameter. The particles are placed in a vacuum chamber and the plant tissue to be engineered is placed below the chamber. The particles are propelled at high velocity using a short pulse of high pressure helium gas, and hit a fine mesh baffle placed above the tissue while the DNA coating continues into any target cell or tissue.

Agrobacterium Method

Transformation via *Agrobacterium* has been successfully practiced in dicots, i.e. broadleaf plants, such as soybeans and tomatoes, for many years. Recently it has been adapted and is now effective in monocots like grasses, including corn and rice. In general, the *Agrobacterium* method is considered preferable to the gene gun, because of a greater frequency of single-site insertions of the foreign DNA, which allows for easier monitoring. In this method, the tumor inducing (Ti) region is removed from the T-DNA (transfer DNA) and replaced with the desired gene and a marker, which is then inserted into the organism. This may involve direct inoculation of the tissue with a culture of transformed Agrobacterium, or inoculation following treatment with micro-projectile bombardment, which wounds the tissue. Wounding of the target tissue causes the release of phenolic compounds by the plant, which induces invasion of the tissue by Agrobacterium. Because of this, microprojectile bombardment often increases the efficiency of infection with Agrobacterium. The marker is used to find the organism which has successfully taken up the desired gene. Tissues of the organism are then transferred to a medium containing an antibiotic or herbicide, depending on which marker was used. The *Agrobacterium* present is also killed by the antibiotic. Only tissues expressing the marker will survive and possess the gene of interest. Thus, subsequent steps in the process will only use these surviving plants. In order to obtain whole plants from these tissues, they are grown under controlled environmental conditions in tissue culture. This is a process of a series of media, each containing nutrients and hormones. Once the plants are grown and produce seed, the process of evaluating the progeny begins. This process entails selection of the seeds with the desired traits and then retesting and growing to make sure that the entire process has been completed successfully with the desired results.

Genetically Engineered Crops

Genetically Engineered Crops

The use of genetically engineered crops has helped many farmers deal with pest problems that reduce their crop production. The impact of pest-resistant crops has led to a much higher yield for farmers in today's world. They can use less pesticides which reduces the chemicals that they put into the ground. Certain engineered crops have led to farmers all over the world and in the United States to increase crop yield exponentially in recent years. Farmers can use a glyphosate herbicide to kill weeds, yet the genetically engineered corn is resistant to the herbicide and is left unaffected. Thus, fields are produced that are virtually weed free. Genetically engineered crops can also benefit farmers when dealing with potentially harmful viruses and bacteria. In the 1990s a mutant strain of virus was decimating the commercial corn fields of the United States. Scientists found a virus resistant strain of maize in the highlands of Mexico and extracted the part of the maize's genome that coded for resistance against the virus and incorporated it into their existing strain of commercial corn. This allowed the commercial strain to produce progeny that were resistant to the virus. Thus, the crops were saved from decimation.

Potential Detrimental Effects of Genetically Engineered Plants

According to John E. Berringer the outcome of releasing genetically modified organisms into the environment is still not known.

2

Key Concepts of Plant Genetics

Plant genetics deals with the heredity of plants. The number of gene characteristics that are present in a species is known as genetic diversity. Topics included in this section include genetic diversity, genetic variability, plant tissue culture, biological hybrid and eugenics. The section strategically encompasses and incorporates the major key concepts of plant genetics, providing a complete understanding.

Genetic Diversity

Genetic diversity is the total number of genetic characteristics in the genetic makeup of a species. It is distinguished from genetic variability, which describes the tendency of genetic characteristics to vary.

Genetic diversity serves as a way for populations to adapt to changing environments. With more variation, it is more likely that some individuals in a population will possess variations of alleles that are suited for the environment. Those individuals are more likely to survive to produce offspring bearing that allele. The population will continue for more generations because of the success of these individuals.

The academic field of population genetics includes several hypotheses and theories regarding genetic diversity. The neutral theory of evolution proposes that diversity is the result of the accumulation of neutral substitutions. Diversifying selection is the hypothesis that two subpopulations of a species live in different environments that select for different alleles at a particular locus. This may occur, for instance, if a species has a large range relative to the mobility of individuals within it. Frequency-dependent selection is the hypothesis that as alleles become more common, they become more vulnerable. This occurs in host–pathogen interactions, where a high frequency of a defensive allele among the host means that it is more likely that a pathogen will spread if it is able to overcome that allele.

Importance of Genetic Diversity

A 2007 study conducted by the National Science Foundation found that genetic diversity and biodiversity (in terms of species diversity) are dependent upon each other—that diversity within a species is necessary to maintain diversity among species, and vice versa. According to the lead researcher in the study, Dr. Richard Lankau, "If any one type is removed from the system, the cycle can break down, and the community becomes dominated by a single species." Genotypic and phenotypic diversity have been found in all species at the protein, DNA, and organismal levels; in nature, this diversity is nonrandom, heavily structured, and correlated with environmental variation and stress.

The interdependence between genetic and species diversity is delicate. Changes in species diversity lead to changes in the environment, leading to adaptation of the remaining species. Changes in genetic diversity, such as in loss of species, leads to a loss of biological diversity. Loss of genetic diversity in domestic animal populations has also been studied and attributed to the extension of markets and economic globalization.

Survival and Adaptation

Genetic diversity plays an important role in the survival and adaptability of a species. When a population's habitat changes, the population may have to adapt to survive; the ability of the population to adapt to the changing environment will determine their ability to cope with an environmental challenge. Variation in the population's gene pool provides variable traits among the individuals of that population. These variable traits can be selected for, via natural selection, ultimately leading to an adaptive change in the population, allowing it to survive in the changed environment. If a population of a species has a very diverse gene pool then there will be more variety in the traits of individuals of that population and consequently more traits for natural selection to act upon to select the fittest individuals to survive.

Genetic diversity is essential for a species to evolve. With very little gene variation within the species, healthy reproduction becomes increasingly difficult, and offspring are more likely to have problems resulting from inbreeding. The vulnerability of a population to certain types of diseases can also increase with reduction in genetic diversity. Concerns about genetic diversity are especially important with large mammals due to their small population size and high levels of human-caused population effects.

Agricultural Relevance

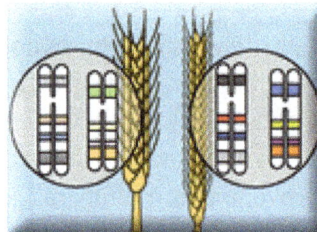

When humans initially started farming, they used selective breeding to pass on desirable traits of the crops while omit-ting the undesirable ones. Selective breeding leads to monocultures: entire farms of nearly genetically identical plants. Little to no genetic diversity makes crops extremely susceptible to widespread disease; bacteria morph and change constantly and when a disease-causing bacterium changes to attack a specific genetic variation, it can easily wipe out vast quantities of the species. If the genetic variation that the bacterium is best at attacking happens to be that which humans have selectively bred to use for harvest, the entire crop will be wiped out.

A very similar occurrence is the cause of the infamous Potato Famine in Ireland. Since new potato plants do not come as a result of reproduction, but rather from pieces of the parent plant, no genetic diversity is developed, and the entire crop is essentially a clone of one potato, it is especially susceptible to an epidemic. In the 1840s, much of Ireland's population depended on potatoes for food. They planted namely the "lumper" variety of potato, which was susceptible to a rot-causing oomycete called *Phytophthora infestans*. This oomycete destroyed the vast majority of the potato crop, and left one million people to starve to death.

Livestock Biodiversity

The genetic diversity of livestock species allows animal production to be practiced in a range of environments and with a range of different objectives. It provides the raw material for selective breeding programmes and allows livestock populations to adapt as environmental conditions change.

Livestock biodiversity can be lost as a result of breed extinctions and other forms of genetic erosion. As of June 2014, among the 8 774 breeds recorded in the Domestic Animal Diversity Information System (DAD-IS), operated by the Food and Agriculture Organization of the United Nations (FAO), 17 percent were classified as being at risk of extinction and 7 percent already extinct. There is now a Global Plan of Action for Animal Genetic Resources that was developed under the auspices of the Commission on Genetic Resources for Food and Agriculture in 2007, that provides a framework and guidelines for the management of animal genetic resources.

Coping with Low Genetic Diversity

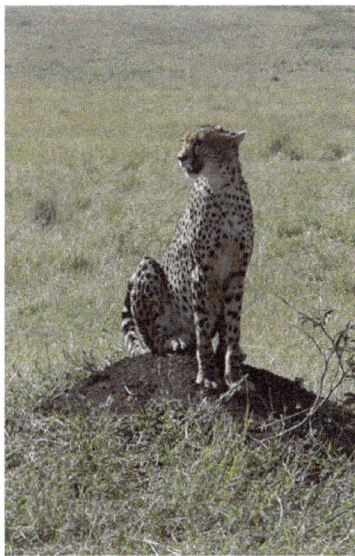

A Tanzanian cheetah.

The natural world has several ways of preserving or increasing genetic diversity. Among oceanic plankton, viruses aid in the genetic shifting process. Ocean viruses, which infect the plankton, carry genes of other organisms in addition to their own. When a virus containing the genes of one cell infects another, the genetic makeup of the latter changes. This constant shift of genetic makeup helps to maintain a healthy population of plankton despite complex and unpredictable environmental changes.

Cheetahs are a threatened species. Low genetic diversity and resulting poor sperm quality has made breeding and survivorship difficult for cheetahs. Moreover, only about 5% of cheetahs survive to adulthood. However, it has been recently discovered that female cheetahs can mate with more than one male per litter of cubs. They undergo induced ovulation, which means that a new egg is produced every time a female mates. By mating with multiple males, the mother increases the genetic diversity within a single litter of cubs.

Measures of Genetic Diversity

Genetic Diversity of a population can be assessed by some simple measures.

- Gene Diversity is the proportion of polymorphic loci across the genome.

- Heterozygosity is the fraction of individuals in a population that are heterozygous for a particular locus.

- Alleles per locus is also used to demonstrate variability.

- Nucleotide diversity is the extent of nucleotide polymorphisms within a population, and is commonly measured through molecular markers such as micro- and minisatellite sequences, mitochondrial DNA, and single-nucleotide polymorphisms (SNPs).

Other Measures of Diversity

Alternatively, other types of diversity may be assessed for organisms:

- species diversity

- ecological diversity

- morphological diversity

- degeneracy

There are broad correlations between different types of diversity. For example, there is a close link between vertebrate taxonomic and ecological diversity.

Genetic Variability

Genetic variability (vary + liable - to or capable of change) is the ability, i.e. capability of a biological system – individual and population – that is changing over time. The base of the genetic variability is genetic variation of different biological systems in space.

Genetic variability is a measure of the tendency of individual genotypes in a population to vary from one another, also. Variability is different from genetic diversity, which is the amount of variation seen in a particular population. The variability of a trait describes how much that trait tends to vary in response to environmental and genetic influences. Genetic variability in a population is important for biodiversity, because without variability, it becomes difficult for a population to adapt to environmental changes and therefore makes it more prone to extinction.

Variability is an important factor in evolution as it affects an individual's response to environmental stress and thus can lead to differential survival of organisms within a population due to natural selection of the most fit variants. Genetic variability also underlies the differential susceptibility of organisms to diseases and sensitivity to toxins or drugs — a fact that has driven increased interest in personalized medicine given the rise of the human genome project and efforts to map the extent of human genetic variation such as the HapMap project.

Causes

There are many sources of genetic variability in a population:

- Homologous recombination is a significant source of variability. During meiosis in sexual organisms, two homologous chromosomes from the male and female parents cross over one another and exchange genetic material. The chromosomes then split apart and are ready to form an offspring. Chromosomal crossover is random and is governed by its own set of genes that code for where crossovers can occur (in cis) and for the mechanism behind the exchange of DNA chunks (in trans). Being controlled by genes means that recombination is also variable in frequency, location, thus it can be selected to increase fitness by nature, because the more recombination the more variability and the more variability the easier it is for the population to handle changes.

- However, recombination during meiosis appears to largely reflect homologous recombinational repair of DNA damages that would otherwise be deleterious to the gametes being produced by meiosis. Thus meiotic processes produce recombinational genetic variation as a byproduct of DNA repair and the level of this variation is related to the level of DNA damaging conditions.

- Immigration, emigration, and translocation – each of these is the movement of an individual into or out of a population. When an individual comes from a previously genetically isolated population into a new one it will increase the genetic variability of the next generation if it reproduces.

- Polyploidy – having more than two homologous chromosomes allows for even more recombination during meiosis allowing for even more genetic variability in one's offspring.

- Diffuse centromeres – in asexual organisms where the offspring is an exact genetic copy of the parent, there are limited sources of genetic variability. One thing that increased variability, however, is having diffused instead of localized centromeres. Being diffused allows the chromatids to split apart in many different ways allowing for chromosome fragmentation and polyploidy creating more variability.

- Genetic mutations – contribute to the genetic variability within a population and can have positive, negative, or neutral effects on a fitness. This variability can be easily propagated throughout a population by natural selection if the mutation increases the affected individual's fitness and its effects will be minimized/hidden if the mutation is deleterious. However, the smaller a population and its genetic variability are, the more likely the recessive/hidden deleterious mutations will show up causing genetic drift.

DNA damages are very frequent, occurring on average more than 60,000 times a day per cell in humans due to metabolic or hydrolytic processes as summarized in DNA damage (naturally occurring). Most DNA damages are accurately repaired by various DNA repair mechanisms. However, some DNA damages remain and give rise to mutations.

It appears that most spontaneously arising mutations result from error prone replication (trans-lesion synthesis) past a DNA damage in the template strand. For example, in yeast more than 60% of spontaneous single-base pair substitutions and deletions are likely caused by translesion synthesis. Another significant source of mutation is an inaccurate DNA repair process, non-homologous end joining, that is often employed in repair of DNA double-strand breaks. Thus it seems that DNA damages are the underlying cause of most spontaneous mutations, either because of error-prone replication past damages or error-prone repair of damages.

Factors that Decrease Genetic Variability

There are many sources that decrease genetic variability in a population:

- Habitat fragmentation describes a discontinuity in an organism's habitat resulting from a geological process or a human-caused event. This action causes a decreased size of the population's habitat and an increased difficulty for emigration and immigration events. These events result in a greater change for factors such as genetic drift to lower genetic diversity.

  - An example of a species disrupted by habitat fragmentation is displayed in Queensland koalas over the past century. After seeing their population's size almost divide in half due to human-caused habitat destruction, the koalas lost their ability to migrate between sub-populations. This event resulted in populations with low genetic diversity, especially those more isolated than others.

- The founder effect is an event that results in populations with low genetic diversity. The founder effect occurs when a population is founded by few individuals, resulting in poor sampling of alleles in a population.

  - An example of a population that experienced the founder effects is displayed with Afognak Island elk population. The population was founded by eight individuals in the early 1900s and quickly grew to a size 1,400. However, even when the population size increased, the genetic variation remained low.

- Climate change is the drastic change in annual weather patterns. These changes in weather patterns can yield negative consequences for genetic diversity. Driving species out of their fundamental niche, climate changes can lower population size and genetic variation drastically.

Plant Tissue Culture

A rose grown from tissue culture.

Plant tissue culture is a collection of techniques used to maintain or grow plant cells, tissues or organs under sterile conditions on a nutrient culture medium of known composition. Plant tissue culture is widely used to produce clones of a plant in a method known as micropropagation. Different techniques in plant tissue culture may offer certain advantages over traditional methods of propagation, including:

- The production of exact copies of plants that produce particularly good flowers, fruits, or have other desirable traits.

- To quickly produce mature plants.

- The production of multiples of plants in the absence of seeds or necessary pollinators to produce seeds.

- The regeneration of whole plants from plant cells that have been genetically modified.

- The production of plants in sterile containers that allows them to be moved with greatly reduced chances of transmitting diseases, pests, and pathogens.

- The production of plants from seeds that otherwise have very low chances of germinating and growing, i.e.: orchids and *Nepenthes*.

- To clean particular plants of viral and other infections and to quickly multiply these plants as 'cleaned stock' for horticulture and agriculture.

Plant tissue culture relies on the fact that many plant cells have the ability to regenerate a whole plant (totipotency). Single cells, plant cells without cell walls (protoplasts), pieces of leaves, stems or roots can often be used to generate a new plant on culture media given the required nutrients and plant hormones.

Techniques

Preparation of plant tissue for tissue culture is performed under aseptic conditions under HEPA filtered air provided by a laminar flow cabinet. Thereafter, the tissue is grown in sterile containers, such as petri dishes or flasks in a growth room with controlled temperature and light intensity. Living plant materials from the environment are naturally contaminated on their surfaces (and sometimes interiors) with microorganisms, so their surfaces are sterilized in chemical solutions (usually alcohol and sodium or calcium hypochlorite) before suitable samples (known as explants) are taken. The sterile explants are then usually placed on the surface of a sterile solid culture medium, but are sometimes placed directly into a sterile liquid medium, particularly when cell suspension cultures are desired. Solid and liquid media are generally composed of inorganic salts plus a few organic nutrients, vitamins and plant hormones. Solid media are prepared from liquid media with the addition of a gelling agent, usually purified agar.

The composition of the medium, particularly the plant hormones and the nitrogen source (nitrate versus ammonium salts or amino acids) have profound effects on the morphology of the tissues that grow from the initial explant. For example, an excess of auxin will often result in a proliferation of roots, while an excess of cytokinin may yield shoots. A balance of both auxin and cytokinin will often produce an unorganised growth of cells, or callus, but the morphology of the outgrowth will depend on the plant species as well as the medium composition. As cultures grow, pieces are typically sliced off and subcultured onto new media to allow for growth or to alter the morphology of the culture. The skill and experience of the tissue culturist are important in judging which pieces to culture and which to discard.

In vitro tissue culture of potato explants

As shoots emerge from a culture, they may be sliced off and rooted with auxin to produce plantlets which, when mature, can be transferred to potting soil for further growth in the greenhouse as normal plants.

Regeneration Pathways

Plant tissue cultures being grown at a USDA seed bank, the National Center for Genetic Resources Preservation.

The specific differences in the regeneration potential of different organs and explants have various explanations. The significant factors include differences in the stage of the cells in the cell cycle, the availability of or ability to transport endogenous growth regulators, and the metabolic capabilities of the cells. The most commonly used tissue explants are the meristematic ends of the plants like the stem tip, auxiliary bud tip and root tip. These tissues have high rates of cell division and either concentrate or produce required growth regulating substances including auxins and cytokinins.

Shoot regeneration efficiency in tissue culture is usually a quantitative trait that often varies between plant species and within a plant species among subspecies, varieties, cultivars, or ecotypes. Therefore, tissue culture regeneration can become complicated especially when many regeneration procedures have to be developed for different genotypes within the same species.

The three common pathways of plant tissue culture regeneration are propagation from preexisting meristems (shoot culture or nodal culture), organogenesis and non-zygotic embryogenesis.

The propagation of shoots or nodal segments is usually performed in four stages for mass production of plantlets through in vitro vegetative multiplication but organogenesis is a common method of micropropagation that involves tissue regeneration of adventitious organs or axillary buds directly or indirectly from the explants. Non-zygotic embryogenesis is a noteworthy developmental pathway that is highly comparable to that of zygotic embryos and it is an important pathway for producing somaclonal variants, developing artificial seeds, and synthesizing metabolites. Due to the single cell origin of non-zygotic embryos, they are preferred in several regeneration systems for micropropagation, ploidy manipulation, gene transfer, and synthetic seed production. Nonetheless, tissue regeneration via organogenesis has also proved to be advantageous for studying regulatory mechanisms of plant development.

Choice of Explant

The tissue obtained from a plant to be cultured is called an explant.

Explants can be taken from different many parts of a plant, including portions of shoots, leaves, stems, flowers, roots, single undifferentiated cells and from many types of mature cells provided are they still contain living cytoplasm and nuclei and are able de-differentiate and resume cell division. This has given rise to the concept of totipotentency of plant cells. However this is not true for all cells or for all plants. In many species explants of various organs vary in their rates of growth and regeneration, while some do not grow at all. The choice of explant material also determines if the plantlets developed via tissue culture are haploid or diploid. Also the risk of microbial contamination is increased with inappropriate explants.

The first method involving the meristems and induction of multiple shoots is the preferred method for the micropropagation industry since the risks of somaclonal variation (genetic variation induced in tissue culture) are minimal when compared to the other two methods. Somatic embryogenesis is a method that has the potential to be several times higher in multiplication rates and is amenable to handling in liquid culture systems like bioreactors.

Some explants, like the root tip, are hard to isolate and are contaminated with soil microflora that become problematic during the tissue culture process. Certain soil microflora can form tight associations with the root systems, or even grow within the root. Soil particles bound to roots are difficult to remove without injury to the roots that then allows microbial attack. These associated microflora will generally overgrow the tissue culture medium before there is significant growth of plant tissue.

Some cultured tissues are slow in their growth. For them there would be two options: (i) Optimizing the culture medium; (ii) Culturing highly responsive tissues or varieties. Necrosis can spoil cultured tissues. Generally, plant varieties differ in susceptibility to tissue culture necrosis. Thus, by culturing highly responsive varieties (or tissues) it can be managed.

Aerial (above soil) explants are also rich in undesirable microflora. However, they are more easily removed from the explant by gentle rinsing, and the remainder usually can be killed by surface sterilization. Most of the surface microflora do not form tight associations with the plant tissue. Such associations can usually be found by visual inspection as a mosaic, de-colorization or localized necrosis on the surface of the explant.

An alternative for obtaining uncontaminated explants is to take explants from seedlings which are aseptically grown from surface-sterilized seeds. The hard surface of the seed is less permeable to penetration of harsh surface sterilizing agents, such as hypochlorite, so the acceptable conditions of sterilization used for seeds can be much more stringent than for vegetative tissues.

Tissue cultured plants are clones. If the original mother plant used to produce the first explants is susceptible to a pathogen or environmental condition, the entire crop would be susceptible to the same problem. Conversely, any positive traits would remain within the line also.

Applications

Plant tissue culture is used widely in the plant sciences, forestry, and in horticulture. Applications include:

- The commercial production of plants used as potting, landscape, and florist subjects, which uses meristem and shoot culture to produce large numbers of identical individuals.

- To conserve rare or endangered plant species.

- A plant breeder may use tissue culture to screen cells rather than plants for advantageous characters, e.g. herbicide resistance/tolerance.

- Large-scale growth of plant cells in liquid culture in bioreactors for production of valuable compounds, like plant-derived secondary metabolites and recombinant proteins used as biopharmaceuticals.

- To cross distantly related species by protoplast fusion and regeneration of the novel hybrid.

- To rapidly study the molecular basis for physiological, biochemical, and reproductive mechanisms in plants, for example in vitro selection for stress tolerant plants.

- To cross-pollinate distantly related species and then tissue culture the resulting embryo which would otherwise normally die (Embryo Rescue).

- For chromosome doubling and induction of polyploidy, for example doubled haploids, tetraploids, and other forms of polyploids. This is usually achieved by application of antimitotic agents such as colchicine or oryzalin.

- As a tissue for transformation, followed by either short-term testing of genetic constructs or regeneration of transgenic plants.

- Certain techniques such as meristem tip culture can be used to produce clean plant material from virused stock, such as potatoes and many species of soft fruit.

- Production of identical sterile hybrid species can be obtained.

Laboratories

Although some growers and nurseries have their own labs for propagating plants by the technique of tissue culture, a number of independent laboratories provide custom propagation services. The Plant Tissue Culture Information Exchange lists many commercial tissue culture labs. Since plant tissue culture is a very labour-intensive process, this would be an important factor in determining which plants would be commercially viable to propagate in a laboratory.

Hybrid (Biology)

Hercules, a "liger", a lion/tiger hybrid.

In biology a hybrid, also known as cross breed, is the result of mixing, through sexual reproduction, two animals or plants of different breeds, varieties, species or genera. Using genetic terminology, it may be defined as follows.

- *Hybrid* generally refers to any offspring resulting from the breeding of two genetically distinct individuals, which usually will result in a high degree of heterozygosity, though hybrid and heterozygous are not, strictly speaking, synonymous.

- a *genetic hybrid* carries two different alleles of the same gene

- a *structural hybrid* results from the fusion of gametes that have differing structure in at least one chromosome, as a result of structural abnormalities

- a *numerical hybrid* results from the fusion of gametes having different haploid numbers of chromosomes

- a *permanent hybrid* is a situation where only the heterozygous genotype occurs, because all homozygous combinations are lethal.

From a taxonomic perspective, hybrid refers to:

- Offspring resulting from the interbreeding between two animal species or plant species.

- Hybrids between different subspecies within a species (such as between the Bengal tiger and Siberian tiger) are known as *intra-specific* hybrids. Hybrids between different species within the same genus (such as between lions and tigers) are sometimes known as *inter-specific* hybrids or crosses. Hybrids between different genera (such as between sheep and goats) are known as *intergeneric* hybrids. Extremely rare *interfamilial* hybrids have been known to occur (such as the guineafowl hybrids). No *interordinal* (between different orders) animal hybrids are known.

- The third type of hybrid consists of crosses between populations, breeds or cultivars within a single species. This meaning is often used in plant and animal breeding, where hybrids are commonly produced and selected, because they have desirable characteristics not found or inconsistently present in the parent individuals or populations.

Terminology

The term hybrid is derived from Latin *hybrida*, meaning the *"offspring of a tame sow and a wild boar"*, *"child of a freeman and slave"*, etc. The term came into popular use in English in the 19th century, though examples of its use have been found from the early 17th century.

There is a popular convention of naming hybrids by forming portmanteau words. The template for this is the naming of tiger-lion hybrids as liger and tigon in the 1920s. This was playfully (but unsystematically) extended to a number of other hybrids, or hypothetical hybrids, such as beefalo (1960s), humanzee (1980s), cama (1998).

Types of Hybrids

Depending on the parents, there are a number of different types of hybrids;

- *Single cross hybrids* — result from the cross between two true breeding organisms and produces an F1 generation called an F1 hybrid (F1 is short for Filial 1, meaning "first offspring"). The cross between two different homozygous lines produces an F1 hybrid that is heterozygous; having two alleles, one contributed by each parent and typically one is dominant and the other recessive. Typically, the F1 generation is also phenotypically homogeneous, producing offspring that are all similar to each other.

- *Double cross hybrids* — result from the cross between two different F1 hybrids.

- *Three-way cross hybrids* — result from the cross between one parent that is an F1 hybrid and the other is from an inbred line.

- *Triple cross hybrids* — result from the crossing of two different three-way cross hybrids.

- *Population hybrids* — result from the crossing of plants or animals in a population with another population. These include crosses between organisms such as interspecific hybrids or crosses between different breeds.

- *Stable hybrid* – a horticultural term which typically refers to an annual plant that, if grown and bred in a small monoculture free of external pollen (e.g., an air-filtered greenhouse) will produce offspring that are "true to type" with respect to phenotype; i.e., a true breeding organism.

- *Hybrid species* – results from hybrid populations evolving reproductive barriers against their parent species through hybrid speciation.

Interspecific Hybrids

Interspecific hybrids are bred by mating two species, normally from within the same genus. The offspring display traits and characteristics of both parents. The offspring of an interspecific cross are very often sterile; thus, hybrid sterility prevents the movement of genes from one species to the other, keeping both species distinct. Sterility is often attributed to the different number of chromosomes the two species have, for example donkeys have 62 chromosomes, while horses have 64 chromosomes, and mules and hinnies have 63 chromosomes. Mules, hinnies, and other normally sterile interspecific hybrids cannot produce viable gametes, because differences in chromosome structure prevent appropriate pairing and segregation during meiosis, meiosis is disrupted, and viable sperm and eggs are not formed. However, fertility in female mules has been reported with a donkey as the father.

Most often other processes occurring in plants and animals keep gametic isolation and species distinction. Species often have different mating or courtship patterns or behaviors, the breeding seasons may be distinct and even if mating does occur antigenic reactions to the sperm of other species prevent fertilization or embryo development. Hybridisation is much more common among organisms that spawn indiscriminately, like soft corals and among plants.

While it is possible to predict the genetic composition of a backcross *on average*, it is not possible to accurately predict the composition of a particular backcrossed individual, due to random segregation of chromosomes. In a species with two pairs of chromosomes, a twice backcrossed individual would be predicted to contain 12.5% of one species' genome (say, species A). However, it may, in fact, still be a 50% hybrid if the chromosomes from species A were lucky in two successive segregations, and meiotic crossovers happened near the telomeres. The chance of this is fairly high: $\left(\dfrac{1}{2}\right)^{(2\times2)} = \dfrac{1}{16}$ (where the "two times two" comes about from two rounds of meiosis with two chromosomes); however, this probability declines markedly with chromosome number and so the actual composition of a hybrid will be increasingly closer to the predicted composition.

Hybrid Species

While not very common, a few animal species have been recognized as being the result of hybridization. The Lonicera fly is an example of a novel animal species that resulted from natural hybridization. The American red wolf appears to be a hybrid species between gray wolf and coyote, although its taxonomic status has been a subject of controversy. The European edible frog appears to be a species, but is actually a semi-permanent hybrid between pool frogs and marsh frogs. The edible frog population is dependent on the presence of at least one of the parents species to be maintained.

Hybrid species of plants are much more common than animals. Many of the crop species are hybrids, and hybridization appears to be an important factor in speciation in some plant groups.

Examples of Hybrid Animals and Animal Populations Derived from Hybrids

Mammals

A "zonkey", a zebra/donkey hybrid.

A "jaglion", a jaguar/lion hybrid.

- Equid hybrids

  - Mule, a cross of female horse and a male donkey.

  - Hinny, a cross between a female donkey and a male horse. Mule and hinny are examples of reciprocal hybrids.

  - Zebroids

    - Zeedonk or Zonkey, a zebra/donkey cross.

    - Zorse, a zebra/horse cross

    - Zony or Zetland, a zebra/pony cross ("zony" is a generic term; "zetland" is specifically a hybrid of the Shetland pony breed with a zebra)

  - hybrid ass, a cross between a donkey and an onager or Asian wild ass.

- Bovid hybrids

  - Dzo, zo or yakow; a cross between a domestic cow/bull and a yak.

  - Beefalo, a cross of an American bison and a domestic cow. This is a fertile breed; this along with genetic evidence has caused them to be recently reclassified into the same genus, *Bos*.

  - Żubroń, a hybrid between wisent (European bison) and domestic cow.

- Sheep-goat hybrid is the cross between a sheep and a goat, which belong to different genera.

- Ursid hybrids, such as the grizzly-polar bear hybrid, occur between black bears, brown bears, and polar bears.

- Felid hybrids

  - Savannah cat are a fertile *breed* developed originally from a cross between the serval [*Leptailurus serval*] and a domestic cat [*Felis catus*].

  - A hybrid between a Bengal tiger and a Siberian tiger is an example of an *intra-specific* hybrid. It also includes the Indochinese tiger, Sumatran tiger too.

  - Pumapards are the hybrid crosses between a puma and a leopard.

  - Ligers and tigons (crosses between a lion and a tiger - the difference in name due to what species the mother and father were - ligers have a lion father and a tiger mother) and other Panthera hybrids such as the lijagulep. Various other wild cat crosses are known involving the lynx, bobcat, leopard, serval, etc.

- Liligers are the hybrid cross between a male lion and a ligress.

- Bengals are a fertile *breed* developed originally from a cross between the Asian leopard cat [*Prionailurus bengalensis*] and the domestic cat [*Felis catus*].

- Fertile canid hybrids occur between coyotes, wolves, dingoes, jackals and domestic dogs.

- Hybrids between black and white rhinoceroses have been recognized.

- Hybrid camel, a cross between a bactrian camel and a dromedary camel

- Cama, a cross between a camel and a llama, also an intergeneric hybrid.

- Wholphin, a fertile but very rare cross between a false killer whale and a bottlenose dolphin.

- At Chester Zoo in the United Kingdom, a cross between an African elephant (male) and an Asian elephant (female). The male calf was named Motty. He died of intestinal infection after twelve days.

- Bornean and Sumatran orangutan hybrids have occurred in captivity.

Birds

A mule, a domestic canary/goldfinch hybrid.

- Hybrids between spotted owls and barred owls

- Cagebird breeders sometimes breed hybrids between species of finch, such as goldfinch × canary. These birds are known as mules.

- The perlin is a peregrine falcon – merlin hybrid.

- Gamebird hybrids, hybrids between gamebirds and domestic fowl, including chickens, guineafowl and peafowl, interfamilial hybrids.

- Numerous macaw hybrids and lovebird hybrids are also known in aviculture.

- Red kite × black kite: five bred unintentionally at a falconry center in England. (It is reported that the black kite (the male) refused female black kites but mated with two female red kites.)

- The mulard duck, hybrid of the domestic pekin duck and domesticated muscovy ducks.

- In Australia, New Zealand and other areas where the Pacific black duck occurs, it is hybridised by the much more aggressive introduced mallard. This is a concern to wildlife authorities throughout the affected area, as it is seen as Genetic pollution of the black duck gene pool.

- Hybridisation in gulls is a reasonably frequent occurrence in the wild.

Reptiles

- Hybrid iguana, a single-cross hybrid resulting from natural interbreeding between male marine iguanas and female land iguanas since the late 2000s.

- Crestoua, a cross between a Rhacodactylus Ciliatus (crested gecko) and a Rhacodactylus Chahoua.

- Colubrid snakes of the tribe Lampropeltini have been shown to produce fertile hybrid offspring.

- Hybridization between the endemic Cuban crocodile (*Crocodilus rhombifer*) and the widely distributed American crocodile (*Crocodilus acutus*) is causing conservation problems for the former species as a threat to its genetic integrity.

- Saltwater crocodiles (*Crocodylus porosus*) have mated with Siamese crocodiles (*Crocodylus siamensis*) in captivity producing offspring which in many cases have grown over 20 feet (6.1 metres) in length. It is likely that wild hybridization occurred historically in parts of southeast Asia.

- Many species of boas and pythons are known to produce hybrids, such as carball (a cross between a ball python and a carpet python) or a bloodball (a cross between a blood python and a ball python) however, most of these only occur in captivity. Contrary to popular belief, boa–python hybrids are not possible due to their differing reproductive functions. Boas only produce hybrids with other species of boas, and pythons only produce hybrids with other species of pythons.

Amphibians

Japanese giant salamanders and Chinese giant salamanders have created hybrids that threaten the survival of Japanese giant salamanders due to the competition for similar resources in Japan.

Fish

- Blood parrot cichlid, which is probably created by crossing a red head cichlid and a Midas cichlid or red devil cichlid

- A group of about 50 hybrids between Australian blacktip shark and the larger common blacktip shark was found by Australia's East Coast in 2012. This is the only known case of hybridization in sharks.

- Silver bream and common bream commonly produce sterile hybrids.

- Tiger muskie is a sterile hybrid between northern pike and muskellunge.

Insects

Killer bees were created during an attempt to breed a strain of bees that would produce more honey and be better adapted to tropical conditions. This was done by crossing a European honey bee and an African bee.

The *Colias eurytheme* and *C. philodice* butterflies have retained enough genetic compatibility to produce viable hybrid offspring.

Hybrid Plants

A sterile hybrid between *Trillium cernuum* and *T. grandiflorum*

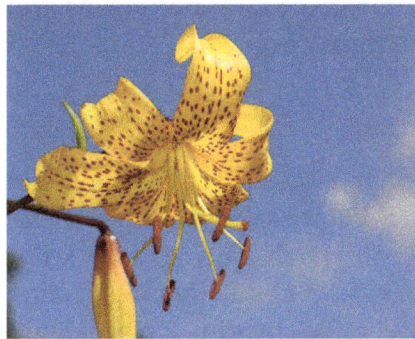

An ornamental lily hybrid known as *Lilium* 'Citronella'

Many hybrids are created by humans, but natural hybrids occur as well. Plant species hybridize more readily than animal species, and the resulting hybrids are more often fertile hybrids and may reproduce, though there still exist sterile hybrids and selective hybrid elimination where the offspring are less able to survive and are thus eliminated before they can reproduce. A number of plant species are the result of hybridization and polyploidy with many plant species easily cross pollinating and producing viable seeds, the distinction between each species is often maintained by geographical isolation or differences in the flowering period. Since plants hybridize frequently without much work, they are often created by humans in order to produce improved plants. These improvements can include the production of more or improved seeds, fruits or other plant parts for consumption, or to make a plant more winter or heat hardy or improve its growth and/or appearance for use in horticulture. Much work is now being done with hybrids to produce more disease resistant plants for both agricultural and horticultural crops. In many groups of plants hybridization has been used to produce larger and more showy flowers and new flower colors. Hybridization may be restricted to the desired parent species through the use of pollination bags.

Many plant genera and species have their origins in polyploidy. Autopolyploidy results from the sudden multiplication in the number of chromosomes in typical normal populations caused by unsuccessful separation of the chromosomes during meiosis. Tetraploids (plants with four sets of chromosomes rather than two) are common in a number of different groups of plants and over time these plants can differentiate into distinct species from the normal diploid line. In *Oenothera lamarchiana* the diploid species has 14 chromosomes, this species has spontaneously given rise to plants with 28 chromosomes that have been given the name *Oenothera gigas*. When hybrids are formed between the tetraploids and the diploid population, the resulting offspring tend to be sterile triploids, thus effectively stopping the intermixing of genes between the two groups of plants (unless the diploids, in rare cases, produce unreduced gametes).

Another form of polyploidy called allopolyploidy occurs when two different species mate and produce polyploid hybrids. Usually the typical chromosome number is doubled, and the four sets of chromosomes can pair up during meiosis, thus the polyploids can produce offspring. Usually, these offspring can mate and reproduce with each other but cannot back-cross with the parent species. Allopolyploids may be able to adapt to new habitats that neither of their parent species inhabited.

Sterility in a non-polyploid hybrid is often a result of chromosome number; if parents are of differing chromosome pair number, the offspring will have an odd number of chromosomes, leaving them unable to produce chromosomally balanced gametes. While this is undesirable in a crop such as wheat, where growing a crop which produces no seeds would be pointless, it is an attractive attribute in some fruits. Triploid bananas and watermelons are intentionally bred because they produce no seeds (and are parthenocarpic).

Heterosis

Hybrids are sometimes stronger than either parent variety, a phenomenon most common with plant hybrids, which when present is known as *hybrid vigor* (heterosis) or heterozygote advantage. A transgressive phenotype is a phenotype displaying more extreme characteristics than either of the parent lines. Plant breeders make use of a number of techniques to produce hybrids, including line breeding and the formation of complex hybrids. An economically important example is hybrid maize (corn), which provides a considerable seed yield advantage over open pollinated varieties. Hybrid seed dominates the commercial maize seed market in the United States, Canada and many other major maize producing countries.

Examples of Plant Hybrids

- The multiplication symbol × (not italicised) indicates a hybrid in the Latin binomial nomenclature. Placed before the binomial it indicates a hybrid between species from different genera (intergeneric hybrid):-

- × *Fatshedera lizei*, a hybrid between *Hedera helix* and *Fatsia japonica*

- × *Heucherella*, a hybrid genus between *Heuchera* and *Tiarella*

- × *Philageria veitchii* is a hybrid between *Lapageria rosea* and *Philesia magellanica*; it is more similar in appearance to the former

- Leyland cypress, [× *Cupressocyparis leylandii*] hybrid between Monterey cypress and

Nootka cypress

- Triticale, [× *Triticosecale*] a wheat–rye hybrid

- × *Urceocharis*, a hybrid between *Eucharis* and *Urceolina*

Interspecific plant hybrids include:

- *Dianthus × allwoodii* (*Dianthus caryophyllus × Dianthus plumarius*)

- Limequat *Citrus × floridana*, key lime *Citrus aurantiifolia* and kumquat *Citrus japonica* hybrid

- Loganberry *Rubus × loganobaccus*, a hybrid between raspberry *Rubus idaeus* and blackberry *Rubus ursinus*

- London plane (*Platanus orientalis × Platanus occidentalis*), thus forming *Platanus × acerifolia*

- *Magnolia × alba* (*Magnolia champaca × Magnolia montana*)

- Peppermint, a hybrid between spearmint and water mint

- *Quercus × warei* (*Quercus robur × Quercus bicolor*) 'Nadler' (marketed in the United States under the trade name Kindred Spirit hybrid oak)

- Tangelo, a hybrid of a Mandarin orange and a pomelo which may have been developed in Asia about 3,500 years ago

- Wheat; most modern and ancient wheat breeds are themselves hybrids. Bread wheat is a hexaploid hybrid of three wild grasses; durum (pasta) wheat is a tetraploid hybrid of two wild grasses

- Grapefruit, hybrid between a pomelo and the Jamaican sweet orange

Some natural hybrids:

- *Iris albicans*, a sterile hybrid which spreads by rhizome division

- Evening primrose, a flower which was the subject of famous experiments by Hugo de Vries on polyploidy and diploidy

Hybrids in Nature

Hybridization between two closely related species is actually a common occurrence in nature but is also being greatly influenced by anthropogenic changes as well. Hybridization is a naturally occurring genetic process where individuals from two genetically distinct populations mate. As stated above, it can occur both intraspecifically, between different distinct populations within the same species, and interspecifically, between two different species. Hybrids can be either sterile/not viable or viable/fertile. This affects the kind of effect that this hybrid will have on its and other populations that it interacts with. Many hybrid zones are known where the ranges of two species meet, and hybrids are continually produced in great numbers. These hybrid zones are useful as biological model systems for studying the mechanisms of speciation (Hybrid speciation). Recently

DNA analysis of a bear shot by a hunter in the North West Territories confirmed the existence of naturally-occurring and fertile grizzly–polar bear hybrids.

Anthropogenic Hybridization

Changes to the environment caused by humans, such as fragmentation and Introduced species, are becoming more widespread. This increases the challenges in managing certain populations that are experiencing introgression, and is a focus of conservation genetics.

Introduced Species and Habitat Fragmentation

Humans have introduced species worldwide to environments for a long time, both intentionally such as establishing a population to be used as a biological control, and unintentionally such as accidental escapes of individuals out of agriculture. This causes drastic global effects on various populations, including through hybridization.

When habitats become broken apart, one of two things can occur, genetically speaking. The first is that populations that were once connected can be cut off from one another, preventing their genes from interacting. Occasionally, this will result in a population of one species breeding with a population of another species as a means of surviving such as the case with the red wolves. Their population numbers being so small, they needed another means of survival. Habitat fragmentation also led to the influx of generalist species into areas where they would not have been, leading to competition and in some cases interbreeding/incorporation of a population into another. In this way, habitat fragmentation is essentially an indirect method of introducing species to an area.

The Hybridization Continuum

There is a kind of continuum with three semi-distinct categories dealing with anthropogenic hybridization: hybridization without Introgression, hybridization with widespread introgression, and essentially a Hybrid swarm. Depending on where a population falls along this continuum, the management plans for that population will change. Hybridization is currently an area of great discussion within Wildlife management and habitat management. Global climate change is creating other changes such as difference in population distributions which are indirect causes for an increase in anthropogenic hybridization.

Consequences

Hybridization can be a less discussed way toward extinction than within detection of where a population lies along the hybrid continuum. The dispute of hybridization is how to manage the resulting hybrids. When a population experiences hybridization with substantial introgression, there still exists parent types of each set of individuals. When a complete hybrid swarm is created, all the individuals are hybrids.

Management of Hybrids

Conservationists disagree on when is the proper time to give up on a population that is becoming a hybrid swarm or to try and save the still existing pure individuals. Once it becomes a complete mixture, we should look to conserve those hybrids to avoid their loss. Most leave it as a case-by-case basis, depending on detecting of hybrids within the group. It is nearly impossible to regulate hybridization via policy because hybridization can occur beneficially when it occurs "naturally" and there is the matter of protecting those previously mentioned hybrid swarms because if they are the only remaining evidence of prior species, they need to be conserved as well.

Expression of Parental Traits in Hybrids

When two distinct types of organisms breed with each other, the resulting hybrids typically have intermediate traits (e.g., one parent has red flowers, the other has white, and the hybrid, pink flowers). Commonly, hybrids also combine traits seen only separately in one parent or the other (e.g., a bird hybrid might combine the yellow head of one parent with the orange belly of the other).

In a hybrid, any trait that falls outside the range of parental variation is termed heterotic. Heterotic hybrids do have new traits, that is, they are not intermediate. *Positive heterosis* produces more robust hybrids, they might be stronger or bigger; while the term *negative heterosis* refers to weaker or smaller hybrids. Heterosis is common in both animal and plant hybrids. For example, hybrids between a lion and a tigress ("ligers") are much larger than either of the two progenitors, while a tigon (lioness × tiger) is smaller. Also the hybrids between the common pheasant (*Phasianus colchicus*) and domestic fowl (*Gallus gallus*) are larger than either of their parents, as are those produced between the common pheasant and hen golden pheasant (*Chrysolophus pictus*). Spurs are absent in hybrids of the former type, although present in both parents.

Genetic Mixing and Extinction

Regionally developed ecotypes can be threatened with extinction when new alleles or genes are introduced that alter that ecotype. This is sometimes called genetic mixing. Hybridization and introgression of new genetic material can lead to the replacement of local genotypes if the hybrids are more fit and have breeding advantages over the indigenous ecotype or species. These hybridization events can result from the introduction of non native genotypes by humans or through habitat modification, bringing previously isolated species into contact. Genetic mixing can be especially detrimental for rare species in isolated habitats, ultimately affecting the population to such a degree that none of the originally genetically distinct population remains.

Effect on Biodiversity and Food Security

In agriculture and animal husbandry, the Green Revolution's use of conventional hybridization increased yields by breeding "high-yielding varieties". The replacement of locally indigenous breeds, compounded with unintentional cross-pollination and crossbreeding (genetic mixing), has reduced the gene pools of various wild and indigenous breeds resulting in the loss of genetic diversity. Since the indigenous breeds are often well-adapted to local extremes in climate and have immunity to local pathogens this can be a significant genetic erosion of the gene pool for future breeding. Therefore, commercial plant geneticists strive to breed "widely adapted" cultivars to counteract this tendency.

Limiting Factors

A number of conditions exist that limit the success of hybridization, the most obvious is great genetic diversity between most species. But in animals and plants that are more closely related hybridization barriers can include morphological differences, differing times of fertility, mating behaviors and cues, physiological rejection of sperm cells or the developing embryo.

In plants, barriers to hybridization include blooming period differences, different pollinator vectors, inhibition of pollen tube growth, somatoplastic sterility, cytoplasmic-genic male sterility and structural differences of the chromosomes.

Mythical, Legendary and Religious Hybrids

Ancient folktales often contain mythological creatures, sometimes these are described as hybrids (e.g., hippogriff as the offspring of a griffin and a horse, and the Minotaur which is the offspring of Pasiphaë and a white bull). More often they are kind of chimera, i.e., a composite of the physical attributes of two or more kinds of animals, mythical beasts, and often humans, with no suggestion that they are the result of interbreeding, e.g., harpies, mermaids, and centaurs.

In the Bible, the Old Testament contains several passages which talk about a first generation of hybrid giants who were known as the Nephilim. The Book of Genesis (6:4) states that "the sons of God went to the daughters of humans and had children by them". As a result, the offspring was born as hybrid giants who became mighty heroes of old and legendary famous figures of ancient times. In addition, the Book of Numbers (13:33) says that the descendants of Anak came from the Nephilim, whose bodies looked exactly like men, but with an enormous height. According to the apocryphal Book of Enoch the Nephilim were wicked sons of fallen angels who had lusted with attractive women.

Eugenics

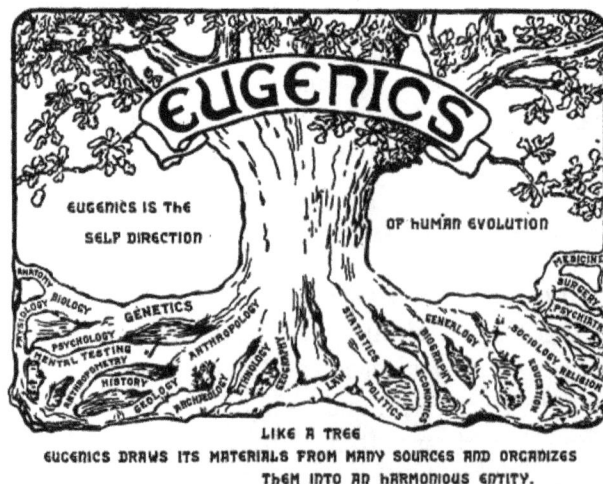

Logo from the Second International Eugenics Conference, 1921, depicting eugenics as a tree which unites a variety of different fields.

Eugenics is a set of beliefs and practices that aims at improving the genetic quality of the human population. It is a social philosophy advocating the improvement of human genetic traits through the promotion of higher rates of sexual reproduction for people with desired traits (positive eugenics), or reduced rates of sexual reproduction and sterilization of people with less-desired or undesired traits (negative eugenics), or both. Alternatively, gene selection rather than "people selection" has recently been made possible through advances in genome editing (e.g. CRISPR). The exact definition of *eugenics* has been a matter of debate since the term was coined. The definition of it as a "social philosophy"—that is, a philosophy with implications for social order—is not universally accepted, and was taken from Frederick Osborn's 1937 journal article "Development of a Eugenic Philosophy".

While eugenic principles have been practiced as far back in world history as Ancient Greece, the modern history of eugenics began in the early 20th century when a popular eugenics movement emerged in the United Kingdom and spread to many countries, including the United States, Canada and most European countries. In this period, eugenic ideas were espoused across the political spectrum. Consequently, many countries adopted eugenic policies meant to improve the genetic stock of their countries. Such programs often included both "positive" measures, such as encouraging individuals deemed particularly "fit" to reproduce, and "negative" measures such as marriage prohibitions and forced sterilization of people deemed unfit for reproduction. People deemed unfit to reproduce often included people with mental or physical disabilities, people who scored in the low ranges of different IQ tests, criminals and deviants, and members of disfavored minority groups. The eugenics movement became negatively associated with Nazi Germany and the Holocaust when many of the defendants at the Nuremberg trials attempted to justify their human rights abuses by claiming there was little difference between the Nazi eugenics programs and the US eugenics programs. In the decades following World War II, with the institution of human rights, many countries gradually abandoned eugenics policies, although some Western countries, among them the United States, continued to carry out forced sterilizations.

Since the 1980s and 1990s when new assisted reproductive technology procedures became available, such as gestational surrogacy (available since 1985), preimplantation genetic diagnosis (available since 1989) and cytoplasmic transfer (first performed in 1996), fear about a possible future revival of eugenics and a widening of the gap between the rich and the poor has emerged.

A major criticism of eugenics policies is that, regardless of whether "negative" or "positive" policies are used, they are vulnerable to abuse because the criteria of selection are determined by whichever group is in political power. Furthermore, negative eugenics in particular is considered by many to be a violation of basic human rights, which include the right to reproduction. Another criticism is that eugenic policies eventually lead to a loss of genetic diversity, resulting in inbreeding depression instead due to a low genetic variation.

Origin and Development

The idea of positive eugenics to produce better human beings has existed at least since Plato suggested selective mating to produce a guardian class. The idea of negative eugenics to decrease the birth of inferior human beings has existed at least since William Goodell (1829-1894) advocated the castration and spaying of the insane.

G. K. Chesterton, an opponent of eugenics, in 1905, by photographer Alvin Langdon Coburn

Francis Galton was an early eugenicist, coining the term itself and popularizing the collocation of the words "nature and nurture".

However, the term "eugenics" to describe a modern project of improving the human population through breeding was originally developed by Francis Galton. Galton had read his half-cousin Charles Darwin's theory of evolution, which sought to explain the development of plant and animal species, and desired to apply it to humans. Based on his biographical studies, Galton believed that desirable human qualities were hereditary traits, though Darwin strongly disagreed with this elaboration of his theory. In 1883, one year after Darwin's death, Galton gave his research a name: *eugenics*. Throughout its recent history, eugenics has remained controversial.

Eugenics became an academic discipline at many colleges and universities, and received funding from many sources. Organisations formed to win public support and sway opinion towards responsible eugenic values in parenthood, including the British Eugenics Education Society of 1907, and the American Eugenics Society of 1921. Both sought support from leading clergymen, and modified their message to meet religious ideals. In 1909 the Anglican clergymen William Inge and James Peile both wrote for the British Eugenics Education Society. Inge was an invited speaker at the 1921 International Eugenics Conference, which was also endorsed by the Roman Catholic Archbishop of New York Patrick Joseph Hayes.

Three International Eugenics Conferences presented a global venue for eugenists with meetings in 1912 in London, and in 1921 and 1932 in New York City. Eugenic policies were first implemented in the early 1900s in the United States. It also took root in France, Germany, and Great Britain. Later, in the 1920s and 30s, the eugenic policy of sterilizing certain mental patients was implemented in other countries, including Belgium, Brazil, Canada, Japan and Sweden.

In addition to being practiced in a number of countries, eugenics was internationally organized through the International Federation of Eugenics Organizations. Its scientific aspects were carried on through research bodies such as the Kaiser Wilhelm Institute of Anthropology, Human Heredity, and Eugenics, the Cold Spring Harbour Carnegie Institution for Experimental Evolution, and the Eugenics Record Office. Politically, the movement advocated measures such as sterilization laws. In its moral dimension, eugenics rejected the doctrine that all human beings are born equal, and redefined moral worth purely in terms of genetic fitness. Its racist elements included pursuit of a pure "Nordic race" or "Aryan" genetic pool and the eventual elimination of "less fit" races.

Early critics of the philosophy of eugenics included the American sociologist Lester Frank Ward, the English writer G. K. Chesterton, the German-American anthropologist Franz Boas, and Scottish tuberculosis pioneer and author Halliday Sutherland. Ward's 1913 article "Eugenics, Euthenics, and Eudemics", Chesterton's 1917 book *Eugenics and Other Evils*, and Boas' 1916 article "Eugenics" (published in *The Scientific Monthly*) were all harshly critical of the rapidly growing movement. Sutherland identified eugenists as a major obstacle to the eradication and cure of tuberculosis in his 1917 address "Consumption: Its Cause and Cure", and criticism of eugenists and Neo-Malthusians in his 1921 book *Birth Control* led to a writ for libel from the eugenist Marie Stopes. Several biologists were also antagonistic to the eugenics movement, including Lancelot Hogben. Other biologists such as J. B. S. Haldane and R. A. Fisher expressed skepticism that sterilization of "defectives" would lead to the disappearance of undesirable genetic traits.

Among institutions, the Catholic Church was an opponent of state-enforced sterilizations. Attempts by the Eugenics Education Society to persuade the British government to legalise voluntary sterilisation were opposed by Catholics and by the Labour Party. The American Eugenics Society initially gained some Catholic supporters, but Catholic support declined following the 1930 papal encyclical *Casti connubii*. In this, Pope Pius XI explicitly condemned sterilization laws: "Public magistrates have no direct power over the bodies of their subjects; therefore, where no crime has taken place and there is no cause present for grave punishment, they can never directly harm, or tamper with the integrity of the body, either for the reasons of eugenics or for any other reason."

As a social movement, eugenics reached its greatest popularity in the early decades of the 20th century, when it was practiced around the world and promoted by governments, institutions, and influential individuals. Many countries enacted various eugenics policies, including: genetic screening, birth control, promoting differential birth rates, marriage restrictions, segregation (both racial segregation and sequestering the mentally ill), compulsory sterilization, forced abortions or forced pregnancies, culminating in genocide.

Nazism and the Decline of Eugenics

Hartheim Euthanasia Centre in 2005

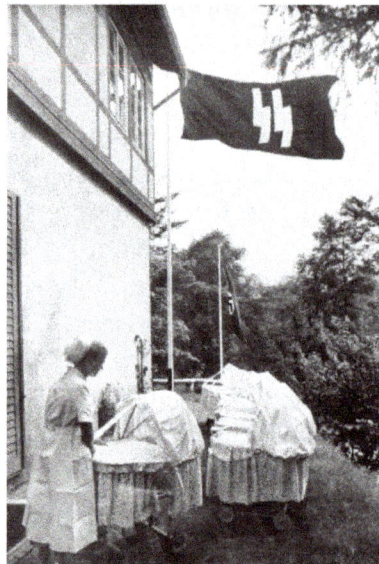

A *Lebensborn* birth house in Nazi Germany. Created with intention of raising the birth rate of "Aryan" children from extramarital relations of "racially pure and healthy" parents.

The scientific reputation of eugenics started to decline in the 1930s, a time when Ernst Rüdin used eugenics as a justification for the racial policies of Nazi Germany. Adolf Hitler had praised and incorporated eugenic ideas in *Mein Kampf* in 1925 and emulated eugenic legislation for the sterilization of "defectives" that had been pioneered in the United States once he took power. Some common early 20th century eugenics methods involved identifying and classifying individuals and their families, including the poor, mentally ill, blind, deaf, developmentally disabled, promiscuous women, homosexuals, and racial groups (such as the Roma and Jews in Nazi Germany) as "degenerate" or "unfit", leading to their their segregation or institutionalization, sterilization, euthanasia, and even their mass murder. The Nazi practice of euthanasia was carried out on hospital patients in the Aktion T4 centers such as Hartheim Castle.

By the end of World War II, many discriminatory eugenics laws were abandoned, having become associated with Nazi Germany. H. G. Wells, who had called for "the sterilization of failures" in 1904, stated in his 1940 book *The Rights of Man: Or What are we fighting for?* that among the human rights he believed should be available to all people was "a prohibition on mutilation, sterilization, torture, and any bodily punishment". After World War II, the practice of "imposing measures intended to prevent births within [a population] group" fell within the definition of the new international crime of genocide, set out in the Convention on the Prevention and Punishment of the Crime of Genocide. The Charter of Fundamental Rights of the European Union also proclaims "the prohibition of eugenic practices, in particular those aiming at selection of persons". In spite of the decline in discriminatory eugenics laws, some government mandated sterilization continued into the 21st century. During the ten years President Alberto Fujimori led Peru from 1990 to 2000, allegedly 2,000 persons were involuntarily sterilized. China maintained its coercive one-child policy until 2015 as well as a suite of other eugenics based legislation to reduce population size and manage fertility rates of different populations. In 2007 the United Nations reported coercive sterilisations and hysterectomies in Uzbekistan. During the years 2005–06 to 2012–13, nearly one-third of the 144 California prison inmates who were sterilized did not give lawful consent to the operation.

Modern Resurgence of Interest

Developments in genetic, genomic, and reproductive technologies at the end of the 20th century are raising numerous questions regarding the ethical status of eugenics, effectively creating a resurgence of interest in the subject. Some, such as UC Berkeley sociologist Troy Duster, claim that modern genetics is a back door to eugenics. This view is shared by White House Assistant Director for Forensic Sciences, Tania Simoncelli, who stated in a 2003 publication by the Population and Development Program at Hampshire College that advances in pre-implantation genetic diagnosis (PGD) are moving society to a "new era of eugenics", and that, unlike the Nazi eugenics, modern eugenics is consumer driven and market based, "where children are increasingly regarded as made-to-order consumer products". In a 2006 newspaper article, Richard Dawkins said that discussion regarding eugenics was inhibited by the shadow of Nazi misuse, to the extent that some scientists would not admit that breeding humans for certain abilities is at all possible. He believes that it is not physically different from breeding domestic animals for traits such as speed or herding skill. Dawkins felt that enough time had elapsed to at least ask just what the ethical differences were between breeding for ability versus training athletes or forcing children to take music lessons, though he could think of persuasive reasons to draw the distinction.

In October 2015, the United Nations' International Bioethics Committee wrote that the ethical problems of human genetic engineering should not be confused with the ethical problems of the 20th century eugenics movements; however, it is still problematic because it challenges the idea of human equality and opens up new forms of discrimination and stigmatization for those who do not want or cannot afford the enhancements.

Transhumanism is often associated with eugenics, although most transhumanists holding similar views nonetheless distance themselves from the term "eugenics" (preferring "germinal choice" or "reprogenetics") to avoid having their position confused with the discredited theories and practices of early-20th-century eugenic movements.

Meanings and Types

Karl Pearson (1912)

The term eugenics and its modern field of study were first formulated by Francis Galton in 1883, drawing on the recent work of his half-cousin Charles Darwin. Galton published his observations and conclusions in his book *Inquiries into Human Faculty and Its Development*.

The origins of the concept began with certain interpretations of Mendelian inheritance, and the theories of August Weismann. The word *eugenics* is derived from the Greek word *eu* ("good" or "well") and the suffix *-genēs* ("born"), and was coined by Galton in 1883 to replace the word "stirpiculture", which he had used previously but which had come to be mocked due to its perceived sexual overtones. Galton defined eugenics as "the study of all agencies under human control which can improve or impair the racial quality of future generations". Galton did not understand the mechanism of inheritance.

Historically, the term has referred to everything from prenatal care for mothers to forced sterilization and euthanasia. To population geneticists, the term has included the avoidance of inbreeding without altering allele frequencies; for example, J. B. S. Haldane wrote that "the motor bus, by breaking up inbred village communities, was a powerful eugenic agent." Debate as to what exactly counts as eugenics has continued to the present day.

Edwin Black, journalist and author of *War Against the Weak*, claims eugenics is often deemed a pseudoscience because what is defined as a genetic improvement of a desired trait is often deemed a cultural choice rather than a matter that can be determined through objective scientific inquiry. The most disputed aspect of eugenics has been the definition of "improvement" of the human gene pool, such as what is a beneficial characteristic and what is a defect. This aspect of eugenics has historically been tainted with scientific racism.

Early eugenists were mostly concerned with perceived intelligence factors that often correlated strongly with social class. Some of these early eugenists include Karl Pearson and Walter Weldon, who worked on this at the University College London.

Eugenics also had a place in medicine. In his lecture "Darwinism, Medical Progress and Eugenics", Karl Pearson said that everything concerning eugenics fell into the field of medicine. He basically placed the two words as equivalents. He was supported in part by the fact that Francis Galton, the father of eugenics, also had medical training.

Eugenic policies have been conceptually divided into two categories. Positive eugenics is aimed at encouraging reproduction among the genetically advantaged; for example, the reproduction of the intelligent, the healthy, and the successful. Possible approaches include financial and political stimuli, targeted demographic analyses, *in vitro* fertilization, egg transplants, and cloning. The movie Gattaca provides a fictional example of positive eugenics done voluntarily. Negative eugenics aimed to eliminate, through sterilization or segregation, those deemed physically, mentally, or morally "undesirable". This includes abortions, sterilization, and other methods of family planning. Both positive and negative eugenics can be coercive; abortion for fit women, for example, was illegal in Nazi Germany.

Jon Entine claims that eugenics simply means "good genes" and using it as synonym for genocide is an "all-too-common distortion of the social history of genetics policy in the United States." According to Entine, eugenics developed out of the Progressive Era and not "Hitler's twisted Final Solution".

Implementation Methods

According to Richard Lynn, eugenics may be divided into two main categories based on the ways in which the methods of eugenics can be applied.

1. Classical Eugenics

    1. Negative eugenics by provision of information and services, i.e. reduction of unplanned pregnancies and births.

        1. "Just say no" campaigns.

        2. Sex education in schools.

        3. School-based clinics.

        4. Promoting the use of contraception.

        5. Emergency contraception.

        6. Research for better contraceptives.

        7. Sterilization.

        8. Abortion.

    2. Negative eugenics by incentives, coercion and compulsion.

        1. Incentives for sterilization.

        2. The Denver Dollar-a-day program, i.e. paying teenage mothers for not becoming pregnant again.

3. Incentives for women on welfare to use contraceptions.

4. Payments for sterilization in developing countries.

5. Curtailment of benefits to welfare mothers.

6. Sterilization of the "mentally retarded".

7. Sterilization of female criminals.

8. Sterilization of male criminals.

3. Licences for parenthood.

4. Positive eugenics.

1. Financial incentives to have children.

2. Selective incentives for childbearing.

3. Taxation of the childless.

4. Ethical obligations of the elite.

5. Eugenic immigration.

2. New Eugenics

1. Artificial insemination by donor.

2. Egg donation.

3. Prenatal diagnosis of genetic disorders and pregnancy terminations of defective fetuses.

4. Embryo selection.

5. Genetic engineering.

6. Gene therapy.

7. Cloning.

Arguments

Efficacy

The first major challenge to conventional eugenics based upon genetic inheritance was made in 1915 by Thomas Hunt Morgan, who demonstrated the event of genetic mutation occurring outside of inheritance involving the discovery of the hatching of a fruit fly (*Drosophila melanogaster*) with white eyes from a family of red-eyes. Morgan claimed that this demonstrated that major genetic changes occurred outside of inheritance and that the concept of eugenics based upon genetic inheritance was not completely scientifically accurate. Additionally, Morgan criticized the view that subjective traits, such as intelligence and criminality, were caused by heredity because he believed that the definitions of these traits varied and that accurate work in genetics could only be done when the traits being studied were accurately defined. In spite of Morgan's public rejection of eugenics, much of his genetic research was absorbed by eugenics.

The heterozygote test is used for the early detection of recessive hereditary diseases, allowing for couples to determine if they are at risk of passing genetic defects to a future child. The goal of the test is to estimate the likelihood of passing the hereditary disease to future descendants.

Recessive traits can be severely reduced, but never eliminated unless the complete genetic make-up of all members of the pool was known, as aforementioned. As only very few undesirable traits, such as Huntington's disease, are dominant, it could be argued from certain perspectives that the practicality of "eliminating" traits is quite low.

There are examples of eugenic acts that managed to lower the prevalence of recessive diseases, although not influencing the prevalence of heterozygote carriers of those diseases. The elevated prevalence of certain genetically transmitted diseases among the Ashkenazi Jewish population (Tay–Sachs, cystic fibrosis, Canavan's disease, and Gaucher's disease), has been decreased in current populations by the application of genetic screening.

Pleiotropy occurs when one gene influences multiple, seemingly unrelated phenotypic traits, an example being phenylketonuria, which is a human disease that affects multiple systems but is caused by one gene defect. Andrzej Pękalski, from the University of Wrocław, argues that eugenics can cause harmful loss of genetic diversity if a eugenics program selects for a pleiotropic gene that is also associated with a positive trait. Pekalski uses the example of a coercive government eugenics program that prohibits people with myopia from breeding but has the unintended consequence of also selecting against high intelligence since the two go together.

Ethics

A common criticism of eugenics is that "it inevitably leads to measures that are unethical". Historically, this statement is evidenced by the obvious control of one group imposing its agenda on minority groups. This includes programs in England, Germany, and America targeting various groups, including Jews, homosexuals, Muslims, Romani, the homeless, and those with intellectual disabilities.

Original position, a hypothetical situation developed by American philosopher John Rawls, has been used as an argument for *negative eugenics*.

Many of the ethical concerns from eugenics arise from the controversial past, prompting a discussion on what place, if any, it should have in the future. Advances in science have changed eugenics. In the past, eugenics has had more to do with sterilization and enforced reproduction laws (i.e. no inter-racial marriage and marriage restrictions based on land ownership). Now, in the age of a progressively mapped genome, embryos can be tested for susceptibility to disease, gender, and genetic defects, and alternative methods of reproduction such as in vitro fertilization are becoming more common. In short, eugenics is no longer *ex post facto* regulation of the living but instead preemptive action on the unborn.

With this change, however, there are ethical concerns which lack adequate attention, and which must be addressed before eugenic policies can be properly implemented in the future. Sterilized individuals, for example, could volunteer for the procedure, albeit under incentive or duress, or at least voice their opinion. The unborn fetus on which these new eugenic procedures are performed cannot speak out, as the fetus lacks the voice to consent or to express his or her opinion. The ability to manipulate a fetus and determine who the child will be is something questioned by many of the opponents of, and even proponents for, eugenic policies.

Societal and political consequences of eugenics call for a place in the discussion on the ethics behind the eugenics movement. Public policy often focuses on issues related to race and gender, both of which could be controlled by manipulation of embryonic genes; eugenics and political issues are interconnected and the political aspect of eugenics must be addressed. Laws controlling the subjects, the methods, and the extent of eugenics will need to be considered in order to prevent the repetition of the unethical events of the past.

Some, such as Nathaniel C. Comfort from Johns Hopkins University, claim that the change from state-led reproductive-genetic decision-making to individual choice has moderated the worst abuses of eugenics by transferring the decision-making from the state to the patient and their family. Comfort suggests that "the eugenic impulse drives us to eliminate disease, live longer and healthier, with greater intelligence, and a better adjustment to the conditions of society; and the health benefits, the intellectual thrill and the profits of genetic bio-medicine are too great for us to do otherwise." Others, such as bioethicist Stephen Wilkinson of Keele University and Honorary Research Fellow Eve Garrard at the University of Manchester, claim that some aspects of modern genetics can be classified as eugenics, but that this classification does not inherently make modern genetics immoral. In a co-authored publication by Keele University, they stated that "[e]ugenics doesn't seem always to be immoral, and so the fact that PGD, and other forms of selective reproduction, might sometimes technically be eugenic, isn't sufficient to show that they're wrong."

 In their 2000 book *From Chance to Choice: Genetics and Justice*, bioethicists Allen Buchanan, Dan Brock, Norman Daniels and Daniel Wikler argued that liberal societies have an obligation to encourage as wide an adoption of eugenic enhancement technologies as possible (so long as such policies do not infringe on individuals' reproductive rights or exert undue pressures on prospective parents to use these technologies) in order to maximize public health and minimize the inequalities that may result from both natural genetic endowments and unequal access to genetic enhancements.

Loss of Genetic Diversity

Eugenic policies could also lead to loss of genetic diversity, in which case a culturally accepted "improvement" of the gene pool could very likely—as evidenced in numerous instances in isolated island populations (e.g., the dodo, *Raphus cucullatus*, of Mauritius)—result in extinction due to increased vulnerability to disease, reduced ability to adapt to environmental change, and other factors both known and unknown. A long-term species-wide eugenics plan might lead to a scenario similar to this because the elimination of traits deemed undesirable would reduce genetic diversity by definition.

Edward M. Miller claims that, in any one generation, any realistic program should make only minor changes in a fraction of the gene pool, giving plenty of time to reverse direction if unintended consequences emerge, reducing the likelihood of the elimination of desirable genes. Miller also argues that any appreciable reduction in diversity is so far in the future that little concern is needed for now.

While the science of genetics has increasingly provided means by which certain characteristics and conditions can be identified and understood, given the complexity of human genetics, culture, and psychology there is at this point no agreed objective means of determining which traits might be ultimately desirable or undesirable. Some diseases such as sickle-cell disease and cystic fibrosis respectively confer immunity to malaria and resistance to cholera when a single copy of the recessive allele is contained within the genotype of the individual. Reducing the instance of sickle-cell disease genes in Africa where malaria is a common and deadly disease could indeed have extremely negative net consequences.

However, some genetic diseases such as haemochromatosis can increase susceptibility to illness, cause physical deformities, and other dysfunctions, which provides some incentive for people to re-consider some elements of eugenics.

Autistic people have advocated a shift in perception of autism spectrum disorders as complex syndromes rather than diseases that must be cured. Proponents of this view reject the notion that there is an "ideal" brain configuration and that any deviation from the norm is pathological; they promote tolerance for what they call neurodiversity. Baron-Cohen argues that the genes for Asperger's combination of abilities have operated throughout recent human evolution and have made remarkable contributions to human history. The possible reduction of autism rates through selection against the genetic predisposition to autism is a significant political issue in the autism rights movement, which claims that autism is a part of neurodiversity.

Many culturally Deaf people oppose attempts to cure deafness, believing instead deafness should be considered a defining cultural characteristic not a disease. Some people have started advocating the idea that deafness brings about certain advantages, often termed "Deaf Gain."

References

- Smith, Andrea (2005). Conquest : sexual violence and American Indian genocide. Cambridge, Massachusetts: South End Press. ISBN 978-0-89608-743-9.

- Ordover, Nancy (2003). American Eugenics: Race, Queer Anatomy, and the Science of Nationalism. Minneapolis: University of Minnesota Press. ISBN 0-8166-3559-5.

- Kerr, Anne; Shakespeare, Tom (2002). Genetic Politics: from Eugenics to Genome. Cheltenham: New Clarion. ISBN 978-1-873797-25-9.

- Joseph, Jay (2006). The Missing Gene: Psychiatry, Heredity, and the Fruitless Search for Genes. New York: Algora. ISBN 978-0-87586-410-5. Archived from the original on 17 April 2009.

- Joseph, Jay (2004). The Gene Illusion: Genetic Research in Psychiatry and Psychology Under the Microscope. New York: Algora. ISBN 978-0-87586-343-6. Archived from the original on 12 May 2009.

- Goldberg, Jonah (2007). Liberal Fascism: The Secret History of the American Left, from Mussolini to the Politics of Meaning (1st ed.). New York: Doubleday. ISBN 0-385-51184-1.

- D'Souza, Dinesh (1995). The End of Racism: Principles for a Multicultural Society. New York: Free Press. ISBN 0-02-908102-5

- Blom, Philipp (2008). The Vertigo Years: Change and Culture in the West, 1900–1914. Toronto: McClelland & Stewart. pp. 335–6. ISBN 978-0-7710-1630-1.

- Gillette, Aaron (2007). The Nature-Nurture Debate in the Twentieth Century. New York: Palgrave Macmillan. ISBN 978-0-230-10845-5.

- Ewen, Elizabeth; Ewen, Stuart (2006). Typecasting: On the Arts and Sciences of Human Inequality (1st ed.). New York: Seven Stories Press. ISBN 978-1-58322-735-0.

- Wyndham, Diana (2003). Eugenics in Australia : striving for national fitness. London: Galton Institute. ISBN 978-0-9504066-7-1.

- Redman, Samuel J. (2016). Bone Rooms: From Scientific Racism to Human Prehistory in Museums. Cambridge: Harvard University Press. ISBN 9780674660410.

3

Genetic Modification in Plants

The crops that are modified by using genetic engineering techniques are called genetically modified plants. It is mostly done to introduce certain traits in plants which are not naturally present in the plant. Some of the key concepts explained in this chapter are genetic variability, plant tissue culture, hybrid and eugenics. The chapter strategically encompasses and incorporates the major components and key concepts of plant genetics, providing a complete understanding.

Genetically Modified Crops

Genetically modified crops (GMCs, GM crops, or biotech crops) are plants used in agriculture, the DNA of which has been modified using genetic engineering techniques. In most cases, the aim is to introduce a new trait to the plant which does not occur naturally in the species. Examples in food crops include resistance to certain pests, diseases, or environmental conditions, reduction of spoilage, or resistance to chemical treatments (e.g. resistance to a herbicide), or improving the nutrient profile of the crop. Examples in non-food crops include production of pharmaceutical agents, biofuels, and other industrially useful goods, as well as for bioremediation.

Farmers have widely adopted GM technology. Between 1996 and 2013, the total surface area of land cultivated with GM crops increased by a factor of 100, from 17,000 km² (4.2 million acres) to 1,750,000 km² (432 million acres). 10% of the world's arable land was planted with GM crops in 2010. In the US, by 2014, 94% of the planted area of soybeans, 96% of cotton and 93% of corn were genetically modified varieties. Use of GM crops expanded rapidly in developing countries, with about 18 million farmers growing 54% of worldwide GM crops by 2013. A 2014 meta-analysis concluded that GM technology adoption had reduced chemical pesticide use by 37%, increased crop yields by 22%, and increased farmer profits by 68%. This reduction in pesticide use has been ecologically beneficial, but benefits may be reduced by overuse. Yield gains and pesticide reductions are larger for insect-resistant crops than for herbicide-tolerant crops. Yield and profit gains are higher in developing countries than in developed countries.

There is a scientific consensus that currently available food derived from GM crops poses no greater risk to human health than conventional food, but that each GM food needs to be tested on a case-by-case basis before introduction. Nonetheless, members of the public are much less likely than scientists to perceive GM foods as safe. The legal and regulatory status of GM foods varies by country, with some nations banning or restricting them, and others permitting them with widely differing degrees of regulation.

However, opponents have objected to GM crops on several grounds, including environmental concerns, whether food produced from GM crops is safe, whether GM crops are needed to address the world's food needs, and concerns raised by the fact these organisms are subject to intellectual property law.

Gene Transfer in Nature and Traditional Agriculture

DNA transfers naturally between organisms. Several natural mechanisms allow gene flow across species. These occur in nature on a large scale – for example, it is one mechanism for the development of antibiotic resistance in bacteria. This is facilitated by transposons, retrotransposons, proviruses and other mobile genetic elements that naturally translocate DNA to new loci in a genome. Movement occurs over an evolutionary time scale.

The introduction of foreign germplasm into crops has been achieved by traditional crop breeders by overcoming species barriers. A hybrid cereal grain was created in 1875, by crossing wheat and rye. Since then important traits including dwarfing genes and rust resistance have been introduced. Plant tissue culture and deliberate mutations have enabled humans to alter the makeup of plant genomes.

History

The first genetically modified crop plant was produced in 1982, an antibiotic-resistant tobacco plant. The first field trials occurred in France and the USA in 1986, when tobacco plants were engineered for herbicide resistance. In 1987, Plant Genetic Systems (Ghent, Belgium), founded by Marc Van Montagu and Jeff Schell, was the first company to genetically engineer insect-resistant (tobacco) plants by incorporating genes that produced insecticidal proteins from *Bacillus thuringiensis* (Bt).

The People's Republic of China was the first country to allow commercialized transgenic plants, introducing a virus-resistant tobacco in 1992, which was withdrawn in 1997. The first genetically modified crop approved for sale in the U.S., in 1994, was the *FlavrSavr* tomato. It had a longer shelf life, because it took longer to soften after ripening. In 1994, the European Union approved tobacco engineered to be resistant to the herbicide bromoxynil, making it the first commercially genetically engineered crop marketed in Europe.

In 1995, Bt Potato was approved by the US Environmental Protection Agency, making it the country's first pesticide producing crop. In 1995 canola with modified oil composition (Calgene), Bt maize (Ciba-Geigy), bromoxynil-resistant cotton (Calgene), Bt cotton (Monsanto), glyphosate-resistant soybeans (Monsanto), virus-resistant squash (Asgrow), and additional delayed ripening tomatoes (DNAP, Zeneca/Peto, and Monsanto) were approved. As of mid-1996, a total of 35 approvals had been granted to commercially grow 8 transgenic crops and one flower crop (carnation), with 8 different traits in 6 countries plus the EU. In 2000, Vitamin A-enriched golden rice was developed, though as of 2016 it was not yet in commercial production. In 2013 the leaders of the three research teams that first applied genetic engineering to crops, Robert Fraley, Marc Van Montagu and Mary-Dell Chilton were awarded the World Food Prize for improving the "quality, quantity or availability" of food in the world.

Methods

Genetically engineered crops have genes added or removed using genetic engineering techniques, originally including gene guns, electroporation, microinjection and agrobacterium. More recently, CRISPR and TALEN offered much more precise and convenient editing techniques.

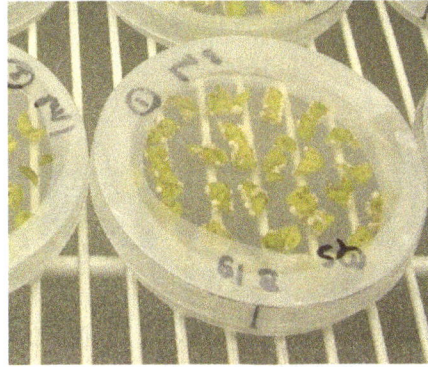

Plants (*Solanum chacoense*) being transformed using agrobacterium

Gene guns (a.k.a. biolistic) "shoot" (direct high energy particles or radiations against) target genes into plant cells. It is the most common method. DNA is bound to tiny particles of gold or tungsten which are subsequently shot into plant tissue or single plant cells under high pressure. The accelerated particles penetrate both the cell wall and membranes. The DNA separates from the metal and is integrated into plant DNA inside the nucleus. This method has been applied successfully for many cultivated crops, especially monocots like wheat or maize, for which transformation using *Agrobacterium tumefaciens* has been less successful. The major disadvantage of this procedure is that serious damage can be done to the cellular tissue.

Agrobacterium tumefaciens-mediated transformation is another common technique. Agrobacteria are natural plant parasites, and their natural ability to transfer genes provides another engineering method. To create a suitable environment for themselves, these Agrobacteria insert their genes into plant hosts, resulting in a proliferation of modified plant cells near the soil level (crown gall). The genetic information for tumour growth is encoded on a mobile, circular DNA fragment (plasmid). When Agrobacterium infects a plant, it transfers this T-DNA to a random site in the plant genome. When used in genetic engineering the bacterial T-DNA is removed from the bacterial plasmid and replaced with the desired foreign gene. The bacterium is a vector, enabling transportation of foreign genes into plants. This method works especially well for dicotyledonous plants like potatoes, tomatoes, and tobacco. Agrobacteria infection is less successful in crops like wheat and maize.

Electroporation is used when the plant tissue does not contain cell walls. In this technique, "DNA enters the plant cells through miniature pores which are temporarily caused by electric pulses."

Microinjection directly injects the gene into the DNA.

Plant scientists, backed by results of modern comprehensive profiling of crop composition, point out that crops modified using GM techniques are less likely to have unintended changes than are conventionally bred crops.

In research tobacco and *Arabidopsis thaliana* are the most frequently modified plants, due to well-developed transformation methods, easy propagation and well studied genomes. They serve as model organisms for other plant species.

Introducing new genes into plants requires a promoter specific to the area where the gene is to be expressed. For instance, to express a gene only in rice grains and not in leaves, an endosperm-specific promoter is used. The codons of the gene must be optimized for the organism due to codon usage bias. Transgenic gene products should be able to be denatured by heat so that they are destroyed during cooking.

Types of Modifications

Transgenic maize containing a gene from the bacteria *Bacillus thuringiensis*

Transgenic

Transgenic plants have genes inserted into them that are derived from another species. The inserted genes can come from species within the same kingdom (plant to plant) or between kingdoms (for example, bacteria to plant). In many cases the inserted DNA has to be modified slightly in order to correctly and efficiently express in the host organism. Transgenic plants are used to express proteins like the cry toxins from *B. thuringiensis*, herbicide resistant genes, antibodies and antigens for vaccinations A study led by the European Food Safety Authority (EFSA) found also viral genes in transgenic plants.

Transgenic carrots have been used to produce the drug Taliglucerase alfa which is used to treat Gaucher's disease. In the laboratory, transgenic plants have been modified to increase photosynthesis (currently about 2% at most plants versus the theoretic potential of 9–10%). This is possible by changing the rubisco enzyme (i.e. changing C3 plants into C4 plants), by placing the rubisco in a carboxysome, by adding CO_2 pumps in the cell wall, by changing the leaf form/size. Plants have been engineered to exhibit bioluminescence that may become a sustainable alternative to electric lighting.

Cisgenic

Cisgenic plants are made using genes found within the same species or a closely related one, where conventional plant breeding can occur. Some breeders and scientists argue that cisgenic modification is useful for plants that are difficult to crossbreed by conventional means (such as potatoes), and that plants in the cisgenic category should not require the same regulatory scrutiny as transgenics.

Subgenic

Genetically modified plants can also be developed using gene knockdown to alter the genetic makeup of a plant without incorporating genes from other plants. In 2014, Chinese researcher Gao Caixia filed patents on the creation of a strain of wheat that is resistant to powdery mildew. The strain lacks genes that encode proteins that repress defenses against the mildew. The researchers deleted all three copies of the genes from wheat's hexaploid genome. Gao used the TALENs and CRISPR gene editing tools without adding or changing any other genes. No field trials were immediately planned. The CRISPR technique has also been used to modify white button mushrooms (*Agaricus bisporus*).

Economics

GM food's economic value to farmers is one of its major benefits, including in developing nations. A 2010 study found that Bt corn provided economic benefits of $6.9 billion over the previous 14 years in five Midwestern states. The majority ($4.3 billion) accrued to farmers producing non-Bt corn. This was attributed to European corn borer populations reduced by exposure to Bt corn, leaving fewer to attack conventional corn nearby. Agriculture economists calculated that "world surplus [increased by] $240.3 million for 1996. Of this total, the largest share (59%) went to U.S. farmers. Seed company Monsanto received the next largest share (21%), followed by US consumers (9%), the rest of the world (6%), and the germplasm supplier, Delta & Pine Land Company of Mississippi (5%)."

According to the International Service for the Acquisition of Agri-biotech Applications (ISAAA), in 2014 approximately 18 million farmers grew biotech crops in 28 countries; about 94% of the farmers were resource-poor in developing countries. 53% of the global biotech crop area of 181.5 million hectares was grown in 20 developing countries. PG Economics comprehensive 2012 study concluded that GM crops increased farm incomes worldwide by $14 billion in 2010, with over half this total going to farmers in developing countries.

Critics challenged the claimed benefits to farmers over the prevalence of biased observers and by the absence of randomized controlled trials. The main Bt crop grown by small farmers in developing countries is cotton. A 2006 review of Bt cotton findings by agricultural economists concluded, "the overall balance sheet, though promising, is mixed. Economic returns are highly variable over years, farm type, and geographical location".

In 2013 the European Academies Science Advisory Council (EASAC) asked the EU to allow the development of agricultural GM technologies to enable more sustainable agriculture, by employing fewer land, water and nutrient resources. EASAC also criticizes the EU's "timeconsuming and expensive regulatory framework" and said that the EU had fallen behind in the adoption of GM technologies.

Participants in agriculture business markets include seed companies, agrochemical companies, distributors, farmers, grain elevators and universities that develop new crops/traits and whose agricultural extensions advise farmers on best practices. According to a 2012 review based on data from the late 1990s and early 2000s, much of the GM crop grown each year is used for livestock feed and increased demand for meat leads to increased demand for GM feedcrops. Feed grain usage as a percentage of total crop production is 70% for corn and more than 90% of oil seed meals such as soybeans. About 65 million metric tons of GM corn grains and about 70 million metric tons of soybean meals derived from GM soybean become feed.

In 2014 the global value of biotech seed was US$15.7 billion; US$11.3 billion (72%) was in industrial countries and US$4.4 billion (28%) was in the developing countries. In 2009, Monsanto had $7.3 billion in sales of seeds and from licensing its technology; DuPont, through its Pioneer subsidiary, was the next biggest company in that market. As of 2009, the overall Roundup line of products including the GM seeds represented about 50% of Monsanto's business.

Some patents on GM traits have expired, allowing the legal development of generic strains that include these traits. For example, generic glyphosate-tolerant GM soybean is now available. Another impact is that traits developed by one vendor can be added to another vendor's proprietary strains, potentially increasing product choice and competition. The patent on the first type of *Roundup Ready* crop that Monsanto produced (soybeans) expired in 2014 and the first harvest of off-patent soybeans occurs in the spring of 2015. Monsanto has broadly licensed the patent to other seed companies that include the glyphosate resistance trait in their seed products. About 150 companies have licensed the technology, including Syngenta and DuPont Pioneer.

Yield

In 2014, the largest review yet concluded that GM crops' effects on farming were positive. The meta-analysis considered all published English-language examinations of the agronomic and economic impacts between 1995 and March 2014 for three major GM crops: soybean, maize, and cotton. The study found that herbicide-tolerant crops have lower production costs, while for insect-resistant crops the reduced pesticide use was offset by higher seed prices, leaving overall production costs about the same.

Yields increased 9% for herbicide tolerance and 25% for insect resistant varieties. Farmers who adopted GM crops made 69% higher profits than those who did not. The review found that GM crops help farmers in developing countries, increasing yields by 14 percentage points.

The researchers considered some studies that were not peer-reviewed, and a few that did not report sample sizes. They attempted to correct for publication bias, by considering sources beyond academic journals. The large data set allowed the study to control for potentially confounding variables such as fertiliser use. Separately, they concluded that the funding source did not influence study results.

Traits

GM crops grown today, or under development, have been modified with various traits. These traits include improved shelf life, disease resistance, stress resistance, herbicide resistance, pest resistance, production of useful goods such as biofuel or drugs, and ability to absorb toxins and for use in bioremediation of pollution.

Recently, research and development has been targeted to enhancement of crops that are locally important in developing countries, such as insect-resistant cowpea for Africa and insect-resistant brinjal (eggplant).

Lifetime

The first genetically modified crop approved for sale in the U.S. was the *FlavrSavr* tomato, which had a longer shelf life. It is no longer on the market.

In November 2014, the USDA approved a GM potato that prevents bruising.

In February 2015 Arctic Apples were approved by the USDA, becoming the first genetically modified apple approved for US sale. Gene silencing was used to reduce the expression of polyphenol oxidase (PPO), thus preventing enzymatic browning of the fruit after it has been sliced open. The trait was added to Granny Smith and Golden Delicious varieties. The trait includes a bacterial antibiotic resistance gene that provides resistance to the antibiotic kanamycin. The genetic engineering involved cultivation in the presence of kanamycin, which allowed only resistant cultivars to survive. Humans consuming apples do not acquire kanamycin resistance, per arcticapple.com. The FDA approved the apples in March 2015.

Nutrition

Edible Oils

Some GM soybeans offer improved oil profiles for processing or healthier eating. Camelina sativa has been modified to produce plants that accumulate high levels of oils similar to fish oils.

Vitamin Enrichment

Golden rice, developed by the International Rice Research Institute (IRRI), provides greater amounts of Vitamin A targeted at reducing Vitamin A deficiency. As of January 2016, golden rice has not yet been grown commercially in any country.

Researchers vitamin-enriched corn derived from South African white corn variety M37W, producing a 169-fold increase in Vitamin A, 6-fold increase in Vitamin C and doubled concentrations of folate. Modified Cavendish bananas express 10-fold the amount of Vitamin A as unmodified varieties.

Toxin Reduction

A genetically modified cassava under development offers lower cyanogen glucosides and enhanced protein and other nutrients (called BioCassava).

In November 2014, the USDA approved a potato, developed by J.R. Simplot Company, that prevents bruising and produces less acrylamide when fried. The modifications prevent natural, harmful proteins from being made via RNA interference. They do not employ genes from non-potato species. The trait was added to the Russet Burbank, Ranger Russet and Atlantic varieties.

Stress Resistance

Plants engineered to tolerate non-biological stressors such as drought, frost, high soil salinity, and nitrogen starvation were in development. In 2011, Monsanto's DroughtGard maize became the first drought-resistant GM crop to receive US marketing approval.

Herbicides

Glyphosate

As of 1999 the most prevalent GM trait was glyphosate-resistance. Glyphosate, (the active ingredient in Roundup and other herbicide products) kills plants by interfering with the shikimate pathway in plants, which is essential for the synthesis of the aromatic amino acids phenylalanine, tyrosine and tryptophan. The shikimate pathway is not present in animals, which instead obtain aromatic amino acids from their diet. More specifically, glyphosate inhibits the enzyme 5-enolpyruvylshikimate-3-phosphate synthase (EPSPS).

This trait was developed because the herbicides used on grain and grass crops at the time were highly toxic and not effective against narrow-leaved weeds. Thus, developing crops that could withstand spraying with glyphosate would both reduce environmental and health risks, and give an agricultural edge to the farmer.

Some micro-organisms have a version of EPSPS that is resistant to glyphosate inhibition. One of these was isolated from an *Agrobacterium* strain CP4 (CP4 EPSPS) that was resistant to glyphosate. The CP4 EPSPS gene was engineered for plant expression by fusing the 5' end of the gene to a chloroplast transit peptide derived from the petunia EPSPS. This transit peptide was used because it had shown previously an ability to deliver bacterial EPSPS to the chloroplasts of other plants. This CP4 EPSPS gene was cloned and transfected into soybeans.

The plasmid used to move the gene into soybeans was PV-GMGTO4. It contained three bacterial genes, two CP4 EPSPS genes, and a gene encoding beta-glucuronidase (GUS) from *Escherichia coli* as a marker. The DNA was injected into the soybeans using the particle acceleration method. Soybean cultivar A5403 was used for the transformation.

Bromoxynil

Tobacco plants have been engineered to be resistant to the herbicide bromoxynil.

Glufosinate

Crops have been commercialized that are resistant to the herbicide glufosinate, as well. Crops engineered for resistance to multiple herbicides to allow farmers to use a mixed group of two, three, or four different chemicals are under development to combat growing herbicide resistance.

2,4-D

In October 2014 the US EPA registered Dow's Enlist Duo maize, which is genetically modified to be resistant to both glyphosate and 2,4-D, in six states. Inserting a bacterial aryloxyalkanoate dioxygenase gene, *aad1* makes the corn resistant to 2,4-D. The USDA had approved maize and soybeans with the mutation in September 2014.

Dicamba

Monsanto has requested approval for a stacked strain that is tolerant of both glyphosate and dicamba.

Pest Resistance

Insects

Tobacco, corn, rice and many other crops have been engineered to express genes encoding for insecticidal proteins from Bacillus thuringiensis (Bt). Papaya, potatoes, and squash have been engineered to resist viral pathogens such as cucumber mosaic virus which, despite its name, infects a wide variety of plants. The introduction of Bt crops during the period between 1996 and 2005 has been estimated to have reduced the total volume of insecticide active ingredient use in the United States by over 100 thousand tons. This represents a 19.4% reduction in insecticide use.

In the late 1990s, a genetically modified potato that was resistant to the Colorado potato beetle was withdrawn because major buyers rejected it, fearing consumer opposition.

Viruses

Virus resistant papaya were developed In response to a papaya ringspot virus (PRV) outbreak in Hawaii in the late 1990s. . They incorporate PRV DNA. By 2010, 80% of Hawaiian papaya plants were genetically modified.

Potatoes were engineered for resistance to potato leaf roll virus and Potato virus Y in 1998. Poor sales led to their market withdrawal after three years.

Yellow squash that were resistant to at first two, then three viruses were developed, beginning in the 1990s. The viruses are watermelon, cucumber and zucchini/courgette yellow mosaic. Squash was the second GM crop to be approved by US regulators. The trait was later added to zucchini.

Many strains of corn have been developed in recent years to combat the spread of Maize dwarf mosaic virus, a costly virus that causes stunted growth which is carried in Johnson grass and spread by aphid insect vectors. These strands are commercially available although the resistance is not standard among GM corn variants.

By-products

Drugs

In 2012, the FDA approved the first plant-produced pharmaceutical, a treatment for Gaucher's Disease. Tobacco plants have been modified to produce therapeutic antibodies.

Biofuel

Algae is under development for use in biofuels. Researchers in Singapore were working on GM jatropha for biofuel production. Syngenta has USDA approval to market a maize trademarked Enogen that has been genetically modified to convert its starch to sugar for ethanol. In 2013, the Flemish Institute for Biotechnology was investigating poplar trees genetically engineered to contain less lignin to ease conversion into ethanol. Lignin is the critical limiting factor when using wood to make bio-ethanol because lignin limits the accessibility of cellulose microfibrils to depolymerization by enzymes.

Materials

Companies and labs are working on plants that can be used to make bioplastics. Potatoes that produce industrially useful starches have been developed as well. Oilseed can be modified to produce fatty acids for detergents, substitute fuels and petrochemicals.

Bioremediation

Scientists at the University of York developed a weed (*Arabidopsis thaliana*) that contains genes from bacteria that can clean TNT and RDX-explosive soil contaminants. 16 million hectares in the

USA (1.5% of the total surface) are estimated to be contaminated with TNT and RDX. However *A. thaliana* was not tough enough for use on military test grounds.

Genetically modified plants have been used for bioremediation of contaminated soils. Mercury, selenium and organic pollutants such as polychlorinated biphenyls (PCBs).

Marine environments are especially vulnerable since pollution such as oil spills are not containable. In addition to anthropogenic pollution, millions of tons of petroleum annually enter the marine environment from natural seepages. Despite its toxicity, a considerable fraction of petroleum oil entering marine systems is eliminated by the hydrocarbon-degrading activities of microbial communities. Particularly successful is a recently discovered group of specialists, the so-called hydrocarbonoclastic bacteria (HCCB) that may offer useful genes.

Asexual Reproduction

Crops such as maize reproduce sexually each year. This randomizes which genes get propagated to the next generation, meaning that desirable traits can be lost. To maintain a high-quality crop, some farmers purchase seeds every year. Typically, the seed company maintains two inbred varieties, and crosses them into a hybrid strain that is then sold. Related plants like sorghum and gamma grass are able to perform apomixis, a form of asexual reproduction that keeps the plant's DNA intact. This trait is apparently controlled by a single dominant gene, but traditional breeding has been unsuccessful in creating asexually-reproducing maize. Genetic engineering offers another route to this goal. Successful modification would allow farmers to replant harvested seeds that retain desirable traits, rather than relying on purchased seed.

Crops

As of 2010 food species for which a genetically modified version is being commercially grown (percent modified in the table below are mostly 2009/2010 data) include:

Crop	Traits	Modification	Percent modified in US	Percent modified in world
Alfalfa	Tolerance of glyphosate or glufosinate	Genes added	Planted in the US from 2005–2007; 2007–2010 court injunction; 2011 approved for sale	
Apples	Delayed browning	Genes added for reduced polyphenol oxidase (PPO) production from other apples	2015 approved for sale	
Canola/ Rapeseed	Tolerance of glyphosate or glufosinate High laurate canola, Oleic acid canola	Genes added	87% (2005)	21%

Corn	Tolerance of herbicides glyphosate glufosinate, and 2,4-D. Insect resistance. Added enzyme, alpha amylase, that converts starch into sugar to facilitate ethanol production. Viral resistance	Genes, some from Bt, added.	Herbicide-resistant: 2013, 85% Bt: 2013, 76% Stacked: 2013, 71%	26%
Cotton (cottonseed oil)	Insect resistance	Genes, some from Bt, added	Herbicide-resistant: 2013, 82% Bt: 2013, 75% Stacked: 2013, 71%	49%
Eggplant	Insect resistance	Genes from Bt	Negligible	Negligible
Papaya (Hawaiian)	Resistance to the papaya ringspot virus.	Gene added	80%	
Potato (food)	Resistance to Colorado beetle Resistance to potato leaf roll virus and Potato virus Y Reduced acrylamide when fried and reduced bruising	Bt cry3A, coat protein from PVY "Innate" potatoes added genetic material coding for mRNA for RNA interference	0%	0%
Potato (starch)	Antibiotic resistance gene, used for selection Better starch production	Antibiotic resistance gene from bacteria Modifications to endogenous starch-producing enzymes	0%	0%
Rice	Enriched with beta-carotene (a source of vitamin A)	Genes from maize and a common soil microorganism.		
Soybeans	Tolerance of glyphosate or glufosinate Reduced saturated fats (high oleic acid); Kills susceptible insect pests Viral resistance	Herbicide resistant gene taken from bacteria added Knocked out native genes that catalyze saturation Gene for one or more Bt crystal proteins added	2014: 94%	77%
Squash (Zucchini/Courgette)	Resistance to watermelon, cucumber and zucchini/courgette yellow mosaic viruses	Viral coat protein genes	13% (figure is from 2005)	
Sugar beet	Tolerance of glyphosate, glufosinate	Genes added	95% (2010); regulated 2011; deregulated 2012	9%

Sugarcane	Pesticide tolerance High sucrose content.	Genes added		
Sweet peppers	Resistance to cucumber mosaic virus	Viral coat protein genes		Small quantities grown in China
Tomatoes	Suppression of the enzyme polygalacturonase (PG), retarding fruit softening after harvesting, while at the same time retaining both the natural color and flavor of the fruit	Antisense gene of the gene responsible for PG enzyme production added	Taken off the market due to commercial failure.	Small quantities grown in China

Development

The number of USDA-approved field releases for testing grew from 4 in 1985 to 1,194 in 2002 and averaged around 800 per year thereafter. The number of sites per release and the number of gene constructs (ways that the gene of interest is packaged together with other elements)—have rapidly increased since 2005. Releases with agronomic properties (such as drought resistance) jumped from 1,043 in 2005 to 5,190 in 2013. As of September 2013, about 7,800 releases had been ap-proved for corn, more than 2,200 for soybeans, more than 1,100 for cotton, and about 900 for po-tatoes. Releases were approved for herbicide tolerance (6,772 releases), insect resistance (4,809), product quality such as flavor or nutrition (4,896), agronomic properties like drought resistance (5,190), and virus/fungal resistance (2,616). The institutions with the most authorized field re-leases include Monsanto with 6,782, Pioneer/ DuPont with 1,405, Syngenta with 565, and USDA's Agricultural Research Service with 370. As of September 2013 USDA had received proposals for releasing GM rice, squash, plum, rose, tobacco, flax and chicory.

Farming Practices

Bt Resistance

Constant exposure to a toxin creates evolutionary pressure for pests resistant to that toxin. Over-reliance on glyphosate and a reduction in the diversity of weed management practices allowed the spread of glyphosate resistance in 14 weed species/biotypes in the US.

One method of reducing resistance is the creation of refuges to allow nonresistant organisms to survive and maintain a susceptible population.

To reduce resistance to Bt crops, the 1996 commercialization of transgenic cotton and maize came with a management strategy to prevent insects from becoming resistant. Insect resistance man-agement plans are mandatory for Bt crops. The aim is to encourage a large population of pests so that any (recessive) resistance genes are diluted within the population. Resistance lowers evolu-tionary fitness in the absence of the stressor (Bt). In refuges, non-resistant strains outcompete resistant ones.

With sufficiently high levels of transgene expression, nearly all of the heterozygotes (S/s), i.e., the largest segment of the pest population carrying a resistance allele, will be killed before maturation, thus preventing transmission of the resistance gene to their progeny. Refuges (i. e., fields of non-transgenic plants) adjacent to transgenic fields increases the likelihood that homozygous resistant (s/s) individuals and any surviving heterozygotes will mate with susceptible (S/S) individuals from the refuge, instead of with other individuals carrying the resistance allele. As a result, the resistance gene frequency in the population remains lower.

Complicating factors can affect the success of the high-dose/refuge strategy. For example, if the temperature is not ideal, thermal stress can lower Bt toxin production and leave the plant more susceptible. More importantly, reduced late-season expression has been documented, possibly resulting from DNA methylation of the promoter. The success of the high-dose/refuge strategy has successfully maintained the value of Bt crops. This success has depended on factors independent of management strategy, including low initial resistance allele frequencies, fitness costs associated with resistance, and the abundance of non-Bt host plants outside the refuges.

Companies that produce Bt seed are introducing strains with multiple Bt proteins. Monsanto did this with Bt cotton in India, where the product was rapidly adopted. Monsanto has also; in an attempt to simplify the process of implementing refuges in fields to comply with Insect Resistance Management(IRM) policies and prevent irresponsible planting practices; begun marketing seed bags with a set proportion of refuge (non-transgenic) seeds mixed in with the Bt seeds being sold. Coined "Refuge-In-a-Bag" (RIB), this practice is intended to increase farmer compliance with refuge requirements and reduce additional labor needed at planting from having separate Bt and refuge seed bags on hand. This strategy is likely to reduce the likelihood of Bt-resistance occurring for corn rootworm, but may increase the risk of resistance for lepidopteran corn pests, such as European corn borer. Increased concerns for resistance with seed mixtures include partially resistant larvae on a Bt plant being able to move to a susceptible plant to survive or cross pollination of refuge pollen on to Bt plants that can lower the amount of Bt expressed in kernels for ear feeding insects.

Herbicide Resistance

Best management practices (BMPs) to control weeds may help delay resistance. BMPs include applying multiple herbicides with different modes of action, rotating crops, planting weed-free seed, scouting fields routinely, cleaning equipment to reduce the transmission of weeds to other fields, and maintaining field borders. The most widely planted GMOs are designed to tolerate herbicides. By 2006 some weed populations had evolved to tolerate some of the same herbicides. Palmer amaranth is a weed that competes with cotton. A native of the southwestern US, it traveled east and was first found resistant to glyphosate in 2006, less than 10 years after GM cotton was introduced.

Plant Protection

Farmers generally use less insecticide when they plant Bt-resistant crops. Insecticide use on corn farms declined from 0.21 pound per planted acre in 1995 to 0.02 pound in 2010. This is consistent with the decline in European corn borer populations as a direct result of Bt corn and cotton. The establishment of minimum refuge requirements helped delay the evolution of Bt resistance. However resistance appears to be developing to some Bt traits in some areas.

Tillage

By leaving at least 30% of crop residue on the soil surface from harvest through planting, conservation tillage reduces soil erosion from wind and water, increases water retention, and reduces soil degradation as well as water and chemical runoff. In addition, conservation tillage reduces the carbon footprint of agriculture. A 2014 review covering 12 states from 1996 to 2006, found that a 1% increase in herbicde-tolerant (HT) soybean adoption leads to a 0.21% increase in conservation tillage and a 0.3% decrease in quality-adjusted herbicide use.

Regulation

The regulation of genetic engineering concerns the approaches taken by governments to assess and manage the risks associated with the development and release of genetically modified crops. There are differences in the regulation of GM crops between countries, with some of the most marked differences occurring between the USA and Europe. Regulation varies in a given country depending on the intended use of each product. For example, a crop not intended for food use is generally not reviewed by authorities responsible for food safety.

Production

In 2013, GM crops were planted in 27 countries; 19 were developing countries and 8 were developed countries. 2013 was the second year in which developing countries grew a majority (54%) of the total GM harvest. 18 million farmers grew GM crops; around 90% were small-holding farmers in developing countries.

Country	2013– GM planted area (million hectares)	Biotech crops
USA	70.1	Maize, Soybean, Cotton, Canola, Sugarbeet, Alfalfa, Papaya, Squash
Brazil	40.3	Soybean, Maize, Cotton
Argentina	24.4	Soybean, Maize, Cotton
India	11.0	Cotton
Canada	10.8	Canola, Maize, Soybean, Sugarbeet
Total	175.2	----

The United States Department of Agriculture (USDA) reports every year on the total area of GMO varieties planted in the United States. According to National Agricultural Statistics Service, the states published in these tables represent 81–86 percent of all corn planted area, 88–90 percent of all soybean planted area, and 81–93 percent of all upland cotton planted area (depending on the year).

Global estimates are produced by the International Service for the Acquisition of Agri-biotech Applications (ISAAA) and can be found in their annual reports, "Global Status of Commercialized Transgenic Crops".

Farmers have widely adopted GM technology. Between 1996 and 2013, the total surface area of land cultivated with GM crops increased by a factor of 100, from 17,000 square kilometers (4,200,000 acres) to 1,750,000 km^2 (432 million acres). 10% of the world's arable land was planted with GM crops in 2010. As of 2011, 11 different transgenic crops were grown commercially on 395 million acres (160 million hectares) in 29 countries such as the USA, Brazil, Argentina, India, Canada, China, Paraguay, Pakistan, South Africa, Uruguay, Bolivia, Australia, Philippines, Myanmar, Burkina Faso, Mexico and Spain. One of the key reasons for this widespread adoption is the perceived economic benefit the technology brings to farmers. For example, the system of planting glyphosate-resistant seed and then applying glyphosate once plants emerged provided farmers with the opportunity to dramatically increase the yield from a given plot of land, since this allowed them to plant rows closer together. Without it, farmers had to plant rows far enough apart to control post-emergent weeds with mechanical tillage. Likewise, using Bt seeds means that farmers do not have to purchase insecticides, and then invest time, fuel, and equipment in applying them. However critics have disputed whether yields are higher and whether chemical use is less, with GM crops.

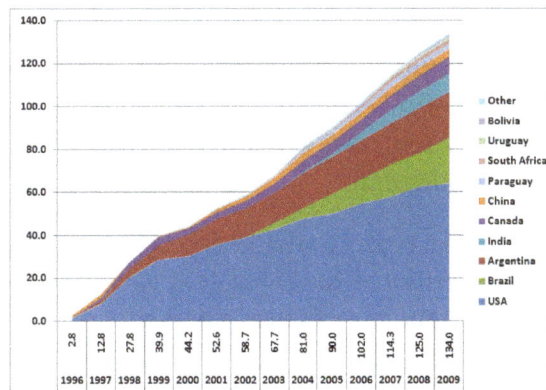

Land area used for genetically modified crops by country (1996–2009), in millions of hectares. In 2011, the land area used was 160 million hectares, or 1.6 million square kilometers.

In the US, by 2014, 94% of the planted area of soybeans, 96% of cotton and 93% of corn were genetically modified varieties. Genetically modified soybeans carried herbicide-tolerant traits only, but maize and cotton carried both herbicide tolerance and insect protection traits (the latter largely Bt protein). These constitute "input-traits" that are aimed to financially benefit the producers, but may have indirect environmental benefits and cost benefits to consumers. The Grocery Manufacturers of America estimated in 2003 that 70–75% of all processed foods in the U.S. contained a GM ingredient.

Europe grows relatively few genetically engineered crops with the exception of Spain, where one fifth of maize is genetically engineered, and smaller amounts in five other countries. The EU had a 'de facto' ban on the approval of new GM crops, from 1999 until 2004. GM crops are now regulated by the EU. In 2015, genetically engineered crops are banned in 38 countries worldwide, 19 of them in Europe. Developing countries grew 54 percent of genetically engineered crops in 2013.

In recent years GM crops expanded rapidly in developing countries. In 2013 approximately 18 million farmers grew 54% of worldwide GM crops in developing countries. 2013's largest increase was in Brazil (403,000 km² versus 368,000 km² in 2012). GM cotton began growing in India in 2002, reaching 110,000 km² in 2013.

According to the 2013 ISAAA brief: "...a total of 36 countries (35 + EU-28) have granted regulatory approvals for biotech crops for food and/or feed use and for environmental release or planting since

1994... a total of 2,833 regulatory approvals involving 27 GM crops and 336 GM events (NB: an "event" is a specific genetic modification in a specific species) have been issued by authorities, of which 1,321 are for food use (direct use or processing), 918 for feed use (direct use or processing) and 599 for environmental release or planting. Japan has the largest number (198), followed by the U.S.A. (165, not including "stacked" events), Canada (146), Mexico (131), South Korea (103), Australia (93), New Zealand (83), European Union (71 including approvals that have expired or under renewal process), Philippines (68), Taiwan (65), Colombia (59), China (55) and South Africa (52). Maize has the largest number (130 events in 27 countries), followed by cotton (49 events in 22 countries), potato (31 events in 10 countries), canola (30 events in 12 countries) and soybean (27 events in 26 countries).

Controversy

GM foods are controversial and the subject of protests, vandalism, referenda, legislation, court action and scientific disputes. The controversies involve consumers, biotechnology companies, governmental regulators, non-governmental organizations and scientists. The key areas are whether GM food should be labeled, the role of government regulators, the effect of GM crops on health and the environment, the effects of pesticide use and resistance, the impact on farmers, and their roles in feeding the world and energy production.

There is a scientific consensus that currently available food derived from GM crops poses no greater risk to human health than conventional food, but that each GM food needs to be tested on a case-by-case basis before introduction. Nonetheless, members of the public are much less likely than scientists to perceive GM foods as safe. The legal and regulatory status of GM foods varies by country, with some nations banning or restricting them, and others permitting them with widely differing degrees of regulation.

No reports of ill effects have been documented in the human population from GM food. Although GMO labeling is required in many countries, the United States Food and Drug Administration does not require labeling, nor does it recognize a distinction between approved GMO and non-GMO foods.

Advocacy groups such as Center for Food Safety, Union of Concerned Scientists, Greenpeace and the World Wildlife Fund claim that risks related to GM food have not been adequately examined and managed, that GMOs are not sufficiently tested and should be labelled, and that regulatory authorities and scientific bodies are too closely tied to industry. Some studies have claimed that genetically modified crops can cause harm; a 2016 review that reanalyzed the data from six of these studies found that their statistical methodologies were flawed and did not demonstrate harm, and said that conclusions about GMO crop safety should be drawn from "the totality of the evidence... instead of far-fetched evidence from single studies".

Advanced Ways to Genetically Modify Crops

Gene Gun

A gene gun or a biolistic particle delivery system, originally designed for plant transformation, is a device for injecting cells with genetic information; the inserted genetic material are termed transgenes. The payload is an elemental particle of a heavy metal coated with plasmid DNA. This technique is often simply referred to as bioballistics or biolistics.

PDS-1000/He Particle Delivery System

This device is able to transform almost any type of cell, including plants, and is not limited to genetic material of the nucleus: it can also transform organelles, including plastids.

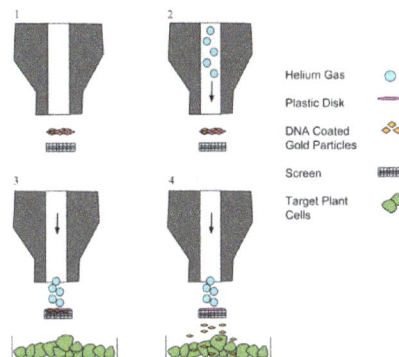

A gene gun is used for injecting cells with genetic information, it is also known as biolistic particle delivery system. Gene guns can be used effectively on most cells but are mainly used on plant cells. Step 1 The gene gun apparatus is ready to fire. Step 2 When the gun is turned on and the helium flows through. Step 3 The helium moving the disk with DNA coated particles toward the screen. Step 4 The helium having pushed the particles moving through the screen and moving to the target cells to transform the cells.

Gene Gun Design

The gene gun was originally a Crosman air pistol modified to fire dense tungsten particles. It was invented by John C Sanford, Ed Wolf and Nelson Allen at Cornell University, and Ted Klein of DuPont, between 1983 and 1986. The original target was onions (chosen for their large cell size) and it was used to deliver particles coated with a marker gene. Genetic transformation was then proven when the onion tissue expressed the gene.

The earliest custom manufactured gene guns (fabricated by Nelson Allen) used a 22 caliber nail gun cartridge to propel an extruded polyethylene cylinder (bullet) down a 22 cal. Douglas barrel. A droplet of the tungsten powder and genetic material was placed on the bullet and shot down the barrel at a lexan "stopping" disk with a petri dish below. The bullet welded to the disk and the genetic information blasted into the sample in the dish with a doughnut effect (devastation in the middle, a ring of good transformation and little around the edge). The gun was connected to a vacuum pump and was under vacuum while firing. The early design was put into limited production by a Rumsey-Loomis (a local machine shop then at Mecklenburg Rd in Ithaca, NY, USA). Later the design was refined by removing the "surge tank" and changing to nonexplosive propellants. DuPont added a plastic extrusion to the exterior to visually improve the machine for mass production to the scientific community. Biorad contracted with Dupont to manufacture and distribute the device. Improvements include the use of helium propellant and a multi-disk-collision delivery mechanism. Other heavy metals such as gold and silver are also used. Gold may be favored because it has better uniformity than tungsten and tungsten can be toxic to cells, but its use may be limited due to availability and cost.

Biolistic Construct Design

A construct is a piece of DNA inserted into the target's genome, including parts that are intended to be removed later. All biolistic transformations require a construct to proceed and while there is great variation among biolistic constructs, they can be broadly sorted into two categories: those which are designed to transform eukaryotic nuclei, and those designed to transform prokaryotic-type genomes such as mitochondria, plasmids or plastids.

Those meant to transform prokaryotic genomes generally have the gene or genes of interest, at least one promoter and terminator sequence, and a reporter gene; which is a gene used to ease detection or removal of those cells which didn't integrate the construct into their DNA. These genes may each have their own promoter and terminator, or be grouped to produce multiple gene products from one transcript, in which case binding sites for translational machinery should be placed between each to ensure maximum translational efficiency. In any case the entire construct is flanked by regions called border sequences which are similar in sequence to locations within the genome, this allows the construct to target itself to a specific point in the existing genome.

Constructs meant for integration into a eukaryotic nucleus follow a similar pattern except that: the construct contains no border sequences because the sequence rearrangement that prokaryotic constructs rely on rarely occurs in eukaryotes; and each gene contained within the construct must be expressed by its own copy of a promoter and terminator sequence.

Though the above designs are generally followed, there are exceptions. For example, the construct might include a Cre-Lox system to selectively remove inserted genes; or a prokaryotic construct may insert itself downstream of a promoter, allowing the inserted genes to be governed by a promoter already in place and eliminating the need for one to be included in the construct.

Application

Gene guns are so far mostly applied for plant cells. However, there is much potential use in humans and other animals as well.

Plants

The target of a gene gun is often a callus of undifferentiated plant cells growing on gel medium in a Petri dish. After the gold particles have impacted the dish, the gel and callus are largely disrupted.

However, some cells were not obliterated in the impact, and have successfully enveloped a DNA coated gold particle, whose DNA eventually migrates to and integrates into a plant chromosome.

Cells from the entire Petri dish can be re-collected and selected for successful integration and expression of new DNA using modern biochemical techniques, such as a using a tandem selectable gene and northern blots.

Selected single cells from the callus can be treated with a series of plant hormones, such as auxins and gibberellins, and each may divide and differentiate into the organized, specialized, tissue cells of an entire plant. This capability of total re-generation is called totipotency. The new plant that originated from a successfully shot cell may have new genetic (heritable) traits.

The use of the gene gun may be contrasted with the use of *Agrobacterium tumefaciens* and its Ti plasmid to insert genetic information into plant cells.

Humans and Other Animals

Gene guns have also been used to deliver DNA vaccines.

The delivery of plasmids into rat neurons through the use of a gene gun, specifically DRG neurons, is also used as a pharmacological precursor in studying the effects of neurodegenerative diseases such as Alzheimer's disease.

The gene gun has become a common tool for labeling subsets of cells in cultured tissue. In addition to being able to transfect cells with DNA plasmids coding for fluorescent proteins, the gene gun can be adapted to deliver a wide variety of vital dyes to cells.

Gene gun bombardment has also been used to transform *Caenorhabditis elegans*, as an alternative to microinjection.

Advantages

Biolistics has proven to be a versatile method of genetic modification and it is generally preferred to engineer transformation-resistant crops, such as cereals. Notably, *Bt* maize is a product of biolistics. Plastid transformation has also seen great success with particle bombardment when compared to other current techniques, such as *Agrobacterium* mediated transformation, which have difficulty targeting the vector to and stably expressing in the chloroplast. In addition, there are no reports of a chloroplast silencing a transgene inserted with a gene gun. Additionally, with only one firing of a gene gun, a skilled technician can generate two transformed organisms. This technology has even allowed for modification of specific tissues *in situ*, although this is likely to damage large numbers of cells and transform only some, rather than all, cells of the tissue.

Limitations

However, biolistics introduces the construct randomly into the target cells. Thus the altered DNA sequences may be transformed into whatever genomes are present in the cell, be they nuclear, mitochondrial, plasmid or any others, in any combination, though proper construct design may mitigate this. Another issue is that the gene inserted may be overexpressed when the construct is inserted multiple times in either the same or different locations of the genome. This is due to the ability of the constructs to give and take genetic information from other constructs, causing some to carry no transgene and others to carry multiple copies; the number of copies inserted depends on both how many copies of the transgene an inserted construct has, and how many were inserted. Also, because eukaryotic constructs rely on illegitimate recombination, a process by which the transgene is integrated into the genome without similar genetic sequences, and not homologous recombination, which inserts at similar sequences, they cannot be targeted to specific locations within the genome.

Agrobacterium

Agrobacterium is a genus of Gram-negative bacteria established by H. J. Conn that uses horizontal gene transfer to cause tumors in plants. *Agrobacterium tumefaciens* is the most commonly studied species in this genus. *Agrobacterium* is well known for its ability to transfer DNA between itself and plants, and for this reason it has become an important tool for genetic engineering.

The *Agrobacterium* genus is quite heterogeneous. Recent taxonomic studies have reclassified all of the *Agrobacterium* species into new genera, such as *Ahrensia*, *Pseudorhodobacter*, *Ruegeria*, and *Stappia*, but most species have been controversially reclassified as *Rhizobium* species.

Plant Pathogen

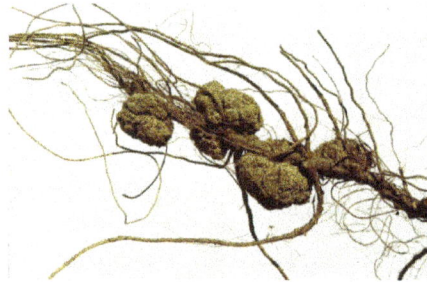

The large growths on these roots are galls induced by *Agrobacterium* sp.

A. tumefaciens causes crown-gall disease in plants. The disease is characterised by a tumour-like growth or gall on the infected plant, often at the junction between the root and the shoot. Tumors are incited by the conjugative transfer of a DNA segment (T-DNA) from the bacterial tumour-inducing (Ti) plasmid. The closely related species, *A. rhizogenes*, induces root tumors, and carries the distinct Ri (root-inducing) plasmid. Although the taxonomy of *Agrobacterium* is currently under revision it can be generalised that 3 biovars exist within the genus, *A. tumefaciens*, *A. rhizogenes*, and *A. vitis*. Strains within *A. tumefaciens* and *A. rhizogenes* are known to be able to harbour either a Ti or Ri-plasmid, whilst strains of *A. vitis*, generally restricted to grapevines, can harbour a Ti-plasmid. Non-*Agrobacterium* strains have been isolated from environmental samples which harbour a Ri-plasmid whilst laboratory studies have shown that non-*Agrobacterium* strains can also harbour a Ti-plasmid. Some environmental strains of *Agrobacterium* possess neither a Ti nor Ri-plasmid. These strains are avirulent.

The plasmid T-DNA is integrated semi-randomly into the genome of the host cell, and the tumor morphology genes on the T-DNA are expressed, causing the formation of a gall. The T-DNA carries genes for the biosynthetic enzymes for the production of unusual amino acids, typically octopine or nopaline. It also carries genes for the biosynthesis of the plant hormones, auxin and cytokinins, and for the biosynthesis of opines, providing a carbon and nitrogen source for the bacteria that most other micro-organisms can't use, giving *Agrobacterium* a selective advantage. By altering the hormone balance in the plant cell, the division of those cells cannot be controlled by the plant, and tumors form. The ratio of auxin to cytokinin produced by the tumor genes determines the morphology of the tumor (root-like, disorganized or shoot-like).

Agrobacterium in Humans

Although generally seen as an infection in plants, *Agrobacterium* can be responsible for opportunistic infections in humans with weakened immune systems, but has not been shown to be a primary pathogen in otherwise healthy individuals. One of the earliest associations of human disease caused by *Agrobacterium radiobacter* was reported by Dr. J. R. Cain in Scotland (1988). A later study suggested that *Agrobacterium* attaches to and genetically transforms several types of human cells by integrating its T-DNA into the human cell genome. The study was conducted using cultured human tissue and did not draw any conclusions regarding related biological activity in nature.

Uses in Biotechnology

The ability of *Agrobacterium* to transfer genes to plants and fungi is used in biotechnology, in particular, genetic engineering for plant improvement. A modified Ti or Ri plasmid can be used. The plasmid is 'disarmed' by deletion of the tumor inducing genes; the only essential parts of the T-DNA are its two small (25 base pair) border repeats, at least one of which is needed for plant transformation. The genes to be introduced into the plant are cloned into a plant transformation vector that contains the T-DNA region of the disarmed plasmid, together with a selectable marker (such as antibiotic resistance) to enable selection for plants that have been successfully transformed. Plants are grown on media containing antibiotic following transformation, and those that do not have the T-DNA integrated into their genome will die. An alternative method is agroinfiltration.

Transformation with *Agrobacterium* can be achieved in two ways. Protoplasts or alternatively leaf-discs can be incubated with the *Agrobacterium* and whole plants regenerated using plant tissue culture. A common transformation protocol for *Arabidopsis* is the floral-dip method: inflorescence are dipped in a suspension of *Agrobacterium*, and the bacterium transforms the germline cells that make the female gametes. The seeds can then be screened for antibiotic resistance (or another marker of interest), and plants that have not integrated the plasmid DNA will die when exposed to the correct condition of antibiotic.

Agrobacterium does not infect all plant species, but there are several other effective techniques for plant transformation including the gene gun.

Agrobacterium is listed as being the vector of genetic material that was transferred to these USA GMOs:

- Soybean
- Cotton
- Corn
- Sugar Beet

- Alfalfa

- Wheat

- Rapeseed Oil (Canola)

- Creeping bentgrass (for animal feed)

- Rice (Golden Rice)

Genomics

The sequencing of the genomes of several species of *Agrobacterium* has permitted the study of the evolutionary history of these organisms and has provided information on the genes and systems involved in pathogenesis, biological control and symbiosis. One important finding is the possibility that chromosomes are evolving from plasmids in many of these bacteria. Another discovery is that the diverse chromosomal structures in this group appear to be capable of supporting both symbiotic and pathogenic lifestyles. The availability of the genome sequences of *Agrobacterium* species will continue to increase, resulting in substantial insights into the function and evolutionary history of this group of plant-associated microbes.

History

Marc Van Montagu and Jozef Schell at the University of Ghent (Belgium) discovered the gene transfer mechanism between *Agrobacterium* and plants, which resulted in the development of methods to alter *Agrobacterium* into an efficient delivery system for gene engineering in plants. A team of researchers led by Dr Mary-Dell Chilton were the first to demonstrate that the virulence genes could be removed without adversely affecting the ability of *Agrobacterium* to insert its own DNA into the plant genome (1983).

List of Genetically Modified Crops

Amflora Potato

Amflora (also known as EH92-527-1) is a genetically modified potato cultivar developed by BASF Plant Science. "Amflora" potato plants produce pure amylopectin starch that is processed to waxy potato starch. It was approved for industrial applications in the European Union on 2 March 2010 by the European Commission. In January 2012, the potato was withdrawn from the market in the EU.

History

Originally registered on 5 August 1996, Amflora was developed by geneticist Lennart Erjefält and agronomist Jüri Känno of Svalöf Weibull AB.

After the European Commission's approval of the potato, BASF announced it was going to produce Amflora seed starting in April 2010 in Germany's Western Pomerania (20 ha) and Sweden (80 ha). It also announced it was planting 150 ha in the Czech Republic "for commercial aims with an unnamed partner."

Due to lack of acceptance of GM crops in Europe, BASF Plant Science decided in January 2012 to stop its commercialization activities in Europe and would no longer sell Amflora there, but it would continue seeking regulatory approval for its products in the Americas and Asia.

In 2013, an EU court annulled the approval of BASF's Amflora, saying that the EU Commission broke rules when it approved the potato in 2010.

Biology

Waxy potato varieties produce two main kinds of potato starch, amylose and amylopectin, the latter of which is most industrially useful. The Amflora potato has been modified to contain antisense against the enzyme that drives synthesis of amylose, namely granule bound starch synthase. This resulting potato almost exclusively produces amylopectin, and thus is more useful for the starch industry.

Industrial Applications

Regular potato starch contains two constituent types of molecules: amylopectin (80 percent), which is more useful as a polymer for industry, and amylose (20 percent) which often creates problems as starch retrogradation, so must be modified with chemical reactions which can be costly.

After two decades of research efforts, BASF's biotechnologists using genetic engineering succeeded in creating a potato, named "Amflora", where the gene responsible for the synthesis of amylose had been turned off, thus the potato is unable to synthesize the less desirable amylose.

Amflora potatoes would be processed and sold as starch to industries that prefer waxy potato starch with only amylopectin. Amflora is intended only for industrial applications such as papermaking and other technical applications. Europe produces more than two million metric tons of natural potato starch a year, and BASF with its Amflora product hoped to enter into this large market.

Other Possible Uses

According to *New York Times*, BASF has a second application pending for use of Amflora's potato pulp as animal feed.

Political Disagreements

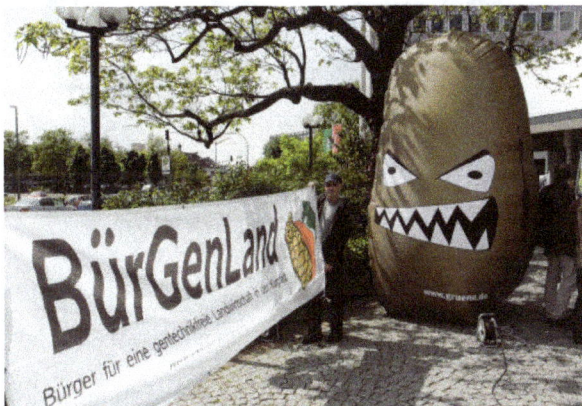

Protests against the Amflora potato

Various environmental organizations, such as Greenpeace, disagreed with the introduction of the Amflora genetically modified potato into the market. The lengthy approval process frustrated some supporters of the potato. A BASF scientist said to the *New York Times*, "it's hard when you see an innovative product go through the loops again and again. These decisions are not about science but about politics". After the potato was approved, the European Greens political party and the Italian agricultural minister criticized the approval. The International Peasant Movement La Via Campesina issued a press release on 8 March 2010 also criticizing the decision.

Reactions by Greek Politicians

After Amflora's licensing by the European Commission on 2 March 2010, the Coalition of the Radical Left's Member of Parliament for the A Thessalonikis prefecture, Tasos Kouvelis, asked the Greek Minister of Agriculture on 3 March 2010 to declare the production of the potato illegal in Greece,. On 4 March 2010 Panhellenic Socialist Movement's European Member of Parliament Kriton Arsenis submitted a question at Europarl asking about the consequences of Amflora.

PASOK's MP Maria Damanaki accepted the decision of the European Commission, while Greek Agriculture Minister Katerina Batzeli said the production of Amflora will not be allowed in Greece.

Licensing Procedure

Amflora could not be sold within the European Union without approval, and its licence could only be issued after voting at the Council of Ministers of the European Union with a 74 percent threshold of support. Two rounds of voting were held, first by experts in December 2006 and then by the agricultural ministers in July 2007, but both failed to reach the 74 percent threshold. Although the voting was by secret ballot, the *New York Times* reported that Amflora was supported by the agricultural ministers of Germany and Belgium, and was opposed by the agricultural ministers of Italy, Ireland, and Austria, while the agricultural ministers of France and Bulgaria abstained from voting.

After a licence was issued on 2 March 2010, BASF announced its intention to ask for approval of more varieties of genetically modified potatoes, such as the "Fortuna" potato.

Golden Rice

Golden rice (right) compared to white rice (left)

Golden rice is a variety of rice (*Oryza sativa*) produced through genetic engineering to biosynthesize beta-carotene, a precursor of vitamin A, in the edible parts of rice. It is intended to produce a fortified food to be grown and consumed in areas with a shortage of dietary vitamin A, a deficiency which is estimated to kill 670,000 children under the age of 5 each year.

Golden rice differs from its parental strain by the addition of three beta-carotene biosynthesis genes. The rice plant can naturally produce beta-carotene in its leaves, where it is involved in photosynthesis. However, the plant does not normally produce the pigment in the endosperm, where photosynthesis does not occur.

In 2005, Golden Rice 2 was announced, which produces up to 23 times more beta-carotene than the original golden rice. To receive the USDA's Recommended Dietary Allowance (RDA), it is estimated that 144 g of the high-yielding strain would have to be eaten. Bioavailability of the carotene from golden rice has been confirmed and found to be an effective source of vitamin A for humans.

Golden rice has met significant opposition from environmental and anti-globalization activists that claim that there are sustainable, long-lasting and more efficient ways to solve vitamin A deficiency that do not compromise food, nutrition and financial security. A study in the Philippines is aimed to evaluate the performance of golden rice, if it can be planted, grown and harvested like other rice varieties, and whether golden rice poses risk to human health. Data has not been released yet.

Golden Rice was one of seven winners of the 2015 Patents for Humanity Awards by the United States Patent and Trademark Office.

As of 2016, it is still in development.

History

The search for a golden rice started off as a Rockefeller Foundation initiative in 1982.

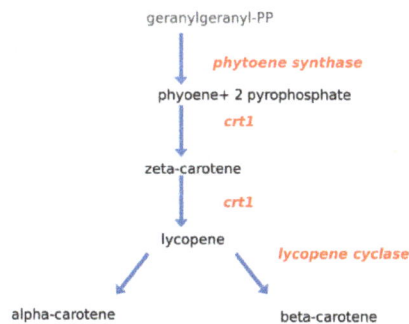

A simplified overview of the carotenoid biosynthesis pathway in golden rice. The enzymes expressed in the endosperm of golden rice, shown in red, catalyze the biosyntheis of beta-carotene from geranylgeranyl diphosphate. Beta-carotene is assumed to be converted to retinal and subsequently retinol (vitamin A) in the animal gut

Peter Bramley discovered in 1990s that a single phytoene desaturase gene (bacterial *CrtI*) can be used to produce lycopene from phytoene in GM tomato, rather than having to introduce multiple carotene desaturases that are normally used by higher plants. Lycopene is then cyclized to beta-carotene by the endogenous cyclase in Golden Rice.

The scientific details of the rice were first published in *Science* in 2000, the product of an eight-year project by Ingo Potrykus of the Swiss Federal Institute of Technology and Peter Beyer of the

University of Freiburg. At the time of publication, golden rice was considered a significant breakthrough in biotechnology, as the researchers had engineered an entire biosynthetic pathway.

The first field trials of golden rice cultivars were conducted by Louisiana State University Agricultural Center in 2004. Additional trials have been conducted in the Philippines and Taiwan, and in Bangladesh (2015). Field testing provides an accurate measurement of nutritional value and enables feeding tests to be performed. Preliminary results from field tests have shown field-grown golden rice produces 4 to 5 times more beta-carotene than golden rice grown under greenhouse conditions.

Crossbreeding

In several countries, Golden rice has been bred with local rice cultivars. or crossbred with the American rice cultivar 'Cocodrie'. As of March 2016, golden rice has not yet been grown commercially, and backcrossing is still ongoing in current varieties to reduce yield drag.

Golden Rice 2

In 2005, a team of researchers at Syngenta produced Golden Rice 2. They combined the phytoene synthase gene from maize with *crt1* from the original golden rice. Golden rice 2 produces 23 times more carotenoids than golden rice (up to 37 µg/g), and preferentially accumulates beta-carotene (up to 31 µg/g of the 37 µg/g of carotenoids).

Genetics

Golden rice was created by transforming rice with two beta-carotene biosynthesis genes:

1. *psy* (phytoene synthase) from daffodil (*Narcissus pseudonarcissus*)

2. *crtI* (carotene desaturase) from the soil bacterium *Erwinia uredovora*

(The insertion of a *lcy* (lycopene cyclase) gene was thought to be needed, but further research showed it is already produced in wild-type rice endosperm.)

The *psy* and *crtI* genes were transferred into the rice nuclear genome and placed under the control of an endosperm-specific promoter, so that they are only expressed in the endosperm. The exogenous *lcy* gene has a transit peptide sequence attached, so it is targeted to the plastid, where geranylgeranyl diphosphate is formed. The bacterial *crtI* gene was an important inclusion to complete the pathway, since it can catalyze multiple steps in the synthesis of carotenoids up to lycopene, while these steps require more than one enzyme in plants. The end product of the engineered pathway is lycopene, but if the plant accumulated lycopene, the rice would be red. Recent analysis has shown the plant's endogenous enzymes process the lycopene to beta-carotene in the endosperm, giving the rice the distinctive yellow color for which it is named. The original golden rice was called SGR1, and under greenhouse conditions it produced 1.6 µg/g of carotenoids.

Vitamin A Deficiency

The research that led to golden rice was conducted with the goal of helping children who suffer from vitamin A deficiency (VAD). In 2005, 190 million children and 19 million pregnant women, in 122 countries, were estimated to be affected by VAD. VAD is responsible for 1–2 million deaths,

500,000 cases of irreversible blindness and millions of cases of xerophthalmia annually. Children and pregnant women are at highest risk. Vitamin A is supplemented orally and by injection in areas where the diet is deficient in vitamin A.

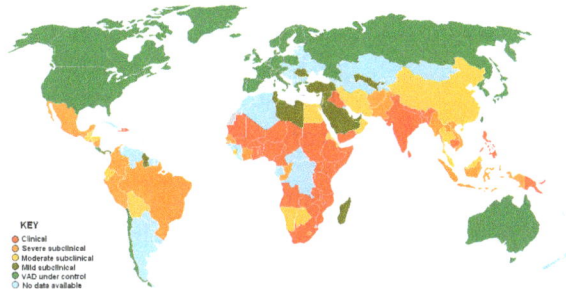

KEY
Clinical
Severe subclinical
Moderate subclinical
Mild subclinical
VAD under control
No data available

Prevalence of vitamin A deficiency. Red is most severe (clinical), green least severe. Countries not reporting data are coded blue. Data collected for a 1995 report.

As of 1999, 43 countries had vitamin A supplementation programs for children under 5; in 10 of these countries, two high dose supplements are available per year, which, according to UNICEF, could effectively eliminate VAD. However, UNICEF and a number of NGOs involved in supplementation note more frequent low-dose supplementation is preferable.

Because many children in VAD-affected countries rely on rice as a staple food, genetic modification to make rice produce the vitamin A precursor beta-carotene was seen as a simple and less expensive alternative to ongoing vitamin supplements or an increase in the consumption of green vegetables or animal products.

Initial analyses of the potential nutritional benefits of golden rice suggested consumption of golden rice would not eliminate the problems of vitamin A deficiency, but could complement other supplementation. Golden Rice 2 contains sufficient provitamin A to provide the entire dietary requirement via daily consumption of some 75g per day.

Since carotenes are hydrophobic, sufficient fat must be present in the diet for golden rice (or most other vitamin A supplements) to alleviate vitamin A deficiency. Vitamin A deficiency is usually coupled to an unbalanced diet. Moreover, this claim referred to an early cultivar of golden rice; one bowl of the latest version provides 60% of RDA for healthy children. The RDA levels advocated in developed countries are far in excess of the amounts needed to prevent blindness.

Research

Dr. José L. Domingo of the Laboratory of Toxicology and Environmental Health, School of Medicine, at Rovira i Virgili University in Spain said, "According to the information reported by the WHO, genetically modified products that are currently on the international market have all passed risk assessments conducted by national authorities." These assessments found no risk to human health. Domingo advocates continued research in the areas of GM rice and its effects on humans.

Clinical Trials/Food Safety and Nutrition Research

In 2009, results of a clinical trial of golden rice with adult volunteers from the US were published in the *American Journal of Clinical Nutrition*. The trial concluded that "beta carotene derived from golden rice is effectively converted to vitamin A in humans". A summary for the American Society for Nutrition suggested that "Golden Rice could probably supply 50% of the Recommended Dietary Allowance (RDA) of vitamin A from a very modest amount — perhaps a cup — of rice, if consumed daily. This amount is well within the consumption habits of most young children and their mothers".

It is well known that beta carotene is found and consumed in many nutritious foods eaten around the world, including fruits and vegetables. Beta carotene in food is a safe source of vitamin A.

The Food Allergy Resource and Research Program of the University of Nebraska undertook research in 2006 that showed the proteins from the new genes in golden rice showed no allergenic properties.

In August 2012, Tufts University and others published research on golden rice in the *American Journal of Clinical Nutrition* showing that the beta carotene produced by golden rice is as effective as beta carotene in oil at providing vitamin A to children. The study stated that "recruitment processes and protocol were approved". In 2015 the journal retracted the study, claiming that the researchers had acted unethically when providing Chinese children golden rice without their parents' consent.

Controversy

Critics of genetically engineered crops have raised various concerns. An early issue was that golden rice originally did not have sufficient vitamin A. This problem was solved by the development of new strains of rice. The speed at which vitamin A degrades once the rice is harvested, and how much remains after cooking are contested. However, a 2009 study concluded that golden rice is effectively converted into vitamin A in humans and a 2012 study that fed 68 children ages 6 to 8 concluded that golden rice was as good as vitamin A supplements and better than the natural beta-carotene in spinach.

Greenpeace opposes the use of any patented genetically modified organisms in agriculture and opposes the cultivation of golden rice, claiming it will open the door to more widespread use of GMOs. The International Rice Research Institute (IRRI) has emphasised the non-commercial nature of their project, stating that "None of the companies listed ... are involved in carrying out the research and development activities of IRRI or its partners in Golden Rice, and none of them will receive any royalty or payment from the marketing or selling of golden rice varieties developed by IRRI."

Vandana Shiva, an Indian anti-GMO activist, argued the problem was not the plant per se, but potential problems with poverty and loss of biodiversity. Shiva claimed these problems could be amplified by the corporate control of agriculture. By focusing on a narrow problem (vitamin A deficiency), Shiva argued, golden rice proponents were obscuring the limited availability of diverse and nutritionally adequate food. Other groups argued that a varied diet containing foods rich in beta carotene such as sweet potato, leaf vegetables and fruit would provide children with sufficient vitamin A. Keith West of Johns Hopkins Bloomberg School of Public Health has stated that foodstuffs containing vitamin A are often unavailable, only available at certain seasons, or too expensive for poor families in underdeveloped countries.

In 2008, WHO malnutrition expert Francesco Branca cited the lack of real-world studies and uncertainty about how many people will use golden rice, concluding "giving out supplements, fortifying existing foods with vitamin A, and teaching people to grow carrots or certain leafy vegetables are, for now, more promising ways to fight the problem".

In 2013, author Michael Pollan, who had critiqued the product in 2001, unimpressed by the benefits, expressed support for the continuation of the research.

Support

The Bill and Melinda Gates Foundation supports the use of genetically modified organisms in agricultural development and supports the International Rice Research Institute in developing Golden Rice.

In June 2016, 107 Nobel laureates signed a letter urging Greenpeace and its supporters to abandon their campaign against GMOs, and against Golden Rice in particular.

Protests

On August 8, 2013 an experimental plot of golden rice being developed at IRRI in the Philippines was uprooted by protesters. British author Mark Lynas reported in Slate that the vandalism was carried out by a group of activists led by the extreme left-inclined *Kilusang Magbubukid ng Pilipinas* (KMP) (unofficial translation: *Farmers' Movement of the Philippines*), to the dismay of other protesters. No local farmers participated in the uprooting; only the small number of activists damaged the golden rice crops because the farmers believe local customs which imply that killing a living rice plant is unlucky.

Distribution

Potrykus has enabled golden rice to be distributed free to subsistence farmers. Free licenses for developing countries were granted quickly due to the positive publicity that golden rice received, particularly in *Time* magazine in July 2000. Monsanto Company was one of the first companies to grant free licences.

The cutoff between humanitarian and commercial use was set at US$10,000. Therefore, as long as a farmer or subsequent user of golden rice genetics does not make more than $10,000 per year, no royalties need to be paid. In addition, farmers are permitted to keep and replant seed.

Bt cotton

Bt cotton is a genetically modified organism (GMO) cotton variety, which produces an insecticide to bollworm. It is produced by Monsanto.

Description

Strains of the bacterium *Bacillus thuringiensis* produce over 200 different Bt toxins, each harmful to different insects. Most notably, Bt toxins are insecticidal to the larvae of moths and butterflies, beetles, cotton bollworms and ghtu flies but are harmless to other forms of life. The gene coding for

Bt toxin has been inserted into cotton as a transgene, causing it to produce this natural insecticide in its tissues. In many regions, the main pests in commercial cotton are lepidopteran larvae, which are killed by the Bt protein in the genetically modified cotton they eat. This eliminates the need to use large amounts of broad-spectrum insecticides to kill lepidopteran pests (some of which have developed pyrethroid resistance). This spares natural insect predators in the farm ecology and further contributes to noninsecticide pest management.

Bt cotton is ineffective against many cotton pests such as plant bugs, stink bugs, and aphids; depending on circumstances it may be desirable to use insecticides in prevention. A 2006 study done by Cornell researchers, the Center for Chinese Agricultural Policy and the Chinese Academy of Science on Bt cotton farming in China found that after seven years these secondary pests that were normally controlled by pesticide had increased, necessitating the use of pesticides at similar levels to non-Bt cotton and causing less profit for farmers because of the extra expense of GM seeds.

Mechanism

Bt cotton was created through the addition of genes encoding toxin crystals in the Cry group of endotoxin. When insects attack and eat the cotton plant the Cry toxins are dissolved due to the high pH level of the insects stomach. The dissolved and activated Cry molecules bond to cadherin-like proteins on cells comprising the brush border molecules. The epithelium of the brush border membranes separates the body cavity from the gut whilst allowing access for nutrients. The Cry toxin molecules attach themselves to specific locations on the cadherin-like proteins present on the epithelial cells of the midge and ion channels are formed which allow the flow of potassium. Regulation of potassium concentration is essential and, if left unchecked, causes death of cells. Due to the formation of Cry ion channels sufficient regulation of potassium ions is lost and results in the death of epithelial cells. The death of such cells creates gaps in the brush border membrane.

History

Bt cotton was first approved for field trials in the United States in 1993, and first approved commercial use in the United States in 1995. Bt cotton was approved by the Chinese government in 1997.

In 2002, a joint venture between Monsanto and Mahyco introduced Bt cotton to India.

In 2011, India grew the largest GM cotton crop at 10.6 million hectares. The U.S. GM cotton crop was 4.0 million hectares, the second largest area in the world, followed by China with 3.9 million hectares and Pakistan with 2.6 million hectares. By 2014, 96% of cotton grown in the United States was genetically modified and 95% of cotton grown in India was GM. India is the largest producer of cotton, and GM cotton, as of 2014.

Advantages

Bt cotton has several advantages over non Bt cotton. The important advantages of Bt cotton are briefly :

- Increases yield of cotton due to effective control of three types of bollworms, viz. American, Spotted and Pink bollworms.

- Insects belonged to Lepidoptera (Bollworms) are sensitive to crystalline endotoxic protein produced by Bt gene which in turn protects cotton from bollworms.

- Reduction in pesticide use in the cultivation of Bt cotton in which bollworms are major pests.

- Reduction in the cost of cultivation and lower farming risks.

- Reduction in environmental pollution by the use of insecticides rarely.

- Bt cotton exhibit genetic resistance or inbuilt resistance which is a permanent type of resistance and not affected by environmental factors. Thus protects crop from bollworms.

- Bt cotton is ecofriendly and does not have adverse effect on parasites, predators, beneficial insecticides and organisms present in soil.

- It promotes multiplication of parasites and predators which help in controlling the bollworms by feeding on larvae and eggs of bollworm.

- No health hazards due to rare use of insecticides (particularly who is engaged in spraying of insecticides).

- Bt cotton are early in maturing as compared to non Bt cotton.

The main selling points of Bt cotton are the reductions in pesticides to be sprayed on a crop and the ecological benefits which stem from that. China first planted Bt cotton in 1997 specifically in response to an outbreak of cotton bollworm, *Helicoverpa armigera*, that farmers were struggling to control with conventional pesticides. Similarly in India and the US, Bt cotton initially alleviated the issues with pests whilst increasing yields and delivering higher profits for farmers.

Studies showed that the lower levels of pesticide being sprayed on the cotton crops promoted biodiversity by allowing non-target species like ladybirds, lacewings and spiders to become more abundant. Likewise it was found that integrated pest management strategies (IPM) were becoming more effective due to the lower levels of pesticide encouraging the growth of natural enemy populations.

Disadvantages

Bt cotton has some limitations

- High cost of Bt cotton seeds as compared to non Bt cotton seeds.

- Effectiveness up to 120 days, after that the toxin producing efficiency of the Bt gene drastically reduces.

- Ineffective against sucking pests like jassids, aphids, whitefly etc.

In India

Bt cotton is supplied in India's Maharashtra state by the agri-biotechnology company Mahyco, which distributes it.

The use of Bt cotton in India has grown exponentially since its introduction in 2002. Recently India has become the number one global exporter of cotton and the second largest cotton producer in the world. India has bred Bt-cotton varieties such as *Bikaneri Nerma* and hybrids such as NHH-44.

Socio-economic surveys confirm that Bt cotton continues to deliver significant and multiple agronomic, economic, environmental and welfare benefits to Indian farmers and society including halved insecticide requirements and a doubling of yields.

India's success has been subject to scrutiny. Monsanto's seeds are expensive and lose vigour after one generation, prompting the Indian Council of Agricultural Research to develop a cheaper Bt cotton variety with seeds that could be reused. The cotton incorporated the cry1Ac gene from the soil bacterium Bacillus thuringiensis (Bt), making the cotton toxic to bollworms. This variety showed poor yield, was removed within a year, and contained a DNA sequence owned by Monsanto, prompting an investigation. In parts of India cases of acquired resistance against Bt cotton have occurred. Monsanto has admitted that the pink bollworm is resistant to first generation transgenic Bt cotton that expresses the single Bt gene (Cry1Ac).

The state of Maharashtra banned the sale and distribution of Bt cotton in 2012, to promote local Indian seeds, which demand less water, fertilizers and pesticide input, but lifted the ban in 2013.

Bt-resistant Pests

After the introduction of Bt cotton in northern China, non-target pests such as mirid bugs (*Heteroptera: Miridae*) became more abundant, because less pesticides were sprayed. In 2013, a second issue being seen across the world, was the development of Bt resistant pests limiting the usefulness of Bt crops.

Main drivers for the widespread resistance in India and China included the high proportion of Bt cotton being planted, 90% and 95% respectively in 2011, and few refuge areas.

Refuge areas of non-Bt crops limit resistance development in targeted pests. The US Environmental Protection Agency requires farmers to have refuge areas of 20%-50% non-Bt crops within 0.8 km of their Bt fields. Such requirements were not seen in China, where instead farmers relied on natural refuge areas to decrease resistance.

In 2009, a novel solution to the resistance problem was trialed in Arizona, when sterile male pink bollworms (*Pectinophora gossypiella*) were released into populations of their wild Bt-resistant counterparts. The hypothesis was that sterile males mating with the few surviving females, who had developed resistance, would lead to a decrease in pests in the following generation. There was a dramatic reduction in pink bollworms, with only two pink bollworm larvae being found by the third year of the study.

Controversies

In India, Bt cotton has been enveloped in controversies due to its supposed links with seed monopolies and farmer suicides. The link between the introduction of Bt cotton to India and a surge in farmer suicides has been refuted by other studies, with decreased farmer suicides since Bt cotton was introduced. Bt cotton accounts for 93% of cotton grown in India.

In the USA

In Hawaii, growing GMO cotton has been prohibited since 2013. Hybridization with the wild cotton species Gossypium tomentosum may be possible. Transgenic cotton is also banned in some parts of Florida.

In Africa

Burkina Faso, Africa's top cotton producer, banned GM cotton in 2016, because of economic and quality concerns.

Flavr Savr

Flavr Savr (also known as CGN-89564-2; pronounced "flavor saver"), a genetically modified tomato, was the first commercially grown genetically engineered food to be granted a license for human consumption. It was produced by the Californian company Calgene, and submitted to the U.S. Food and Drug Administration (FDA) in 1992. On May 18, 1994, the FDA completed its evaluation of the Flavr Savr tomato and the use of APH(3')II, concluding that the tomato "is as safe as tomatoes bred by conventional means" and "that the use of aminoglycoside 3'-phosphotransferase II is safe for use as a processing aid in the development of new varieties of tomato, rapeseed oil, and cotton intended for food use." It was first sold in 1994, and was only available for a few years before production ceased in 1997. Calgene made history, but mounting costs prevented the company from becoming profitable, and it was eventually acquired by Monsanto Company.

Characteristics

Through genetic engineering, Calgene hoped to slow down the ripening process of the tomato and thus prevent it from softening, while still allowing the tomato to retain its natural colour and flavor. The tomato was made more resistant to rotting by adding an antisense gene which interferes with the production of the enzyme polygalacturonase. The enzyme normally degrades pectin in the cell walls and results in the softening of fruit which makes them more susceptible to being damaged by fungal infections. Some unmodified tomatoes are picked before fully ripened and are then artificially ripened using ethylene gas which acts as a plant hormone. Picking the fruit while unripe allows for easier handling and extended shelf-life. Flavr Savr tomatoes, on the other hand, could be allowed to ripen on the vine, without compromising their shelf-life. The intended effect of slowing down the softening of Flavr Savr tomatoes would allow the vine-ripe fruits to be harvested like green tomatoes without greater damage to the tomato itself. The Flavr Savr turned out to disappoint researchers in that respect, as the antisensed PG gene had a positive effect on shelf life, but not on the fruit's firmness, so the tomatoes still had to be harvested like any other unmodified vine-ripe tomatoes. An improved flavor, later achieved through traditional breeding of Flavr Savr and better tasting varieties, would also contribute to selling Flavr Savr at a premium price at the supermarket.

The FDA stated that special labeling for these modified tomatoes was not necessary because they have the essential characteristics of non-modified tomatoes. Specifically, there was no evidence for health risks, and the nutritional content was unchanged.

The failure of the Flavr Savr has been attributed to Calgene's inexperience in the business of growing and shipping tomatoes.

Tomato Paste

In the UK, Zeneca produced a tomato paste that used technology similar to the Flavr Savr. Don Grierson was involved in the research to make the genetically modified tomato. Due to the characteristics of the tomato, it was cheaper to produce than conventional tomato paste, resulting in the product being 20% cheaper. Between 1996 and 1999, 1.8 million cans, clearly labelled as genetically engineered, were sold in Sainsbury's and Safeway. At one point the paste outsold normal tomato paste but sales fell in the autumn of 1998.

The House of Commons of the United Kingdom published a report in which they stated that the decline in sales during this period was linked to changing consumer perceptions of genetically modified crops. The report identified several possible factors, including product labeling and perception of choice, lobbying campaigns, and media attention. It concluded that the tone of media reports on the subject underwent a "fundamental shift" in response to a high profile incident in which Dr. Arpad Pusztai, a researcher for Rowett Research Institute, was fired after making a televised claim about detrimental health effects in lab rats fed a diet of genetically modified potatoes. Subsequent peer review and testimony by Dr. Pusztai led the House Science and Technology Select Committee to conclude that his initial claim was "contradicted by his own evidence." In the intervening period, Sainsbury's and Safeway both pledged that none of their house brand products would contain genetically modified ingredients.

Genetically Modified Maize

Transgenic maize containing a gene from the bacteria *Bacillus thuringiensis*

Genetically modified maize (corn) is a genetically modified crop. Specific maize strains have been genetically engineered to express agriculturally-desirable traits, including resistance to pests and to herbicides. Maize strains with both traits are now in use in multiple countries. GM maize has also caused controversy with respect to possible health effects, impact on other insects and impact on other plants via gene flow. One strain, called Starlink, was approved only for animal feed in the US, but was found in food, leading to a series of recalls starting in 2000.

Marketed Products

Herbicide Resistant Maize

Corn varieties resistant to glyphosate herbicides were first commercialized in 1996 by Monsanto, and are known as "Roundup Ready Corn". They tolerate the use of Roundup. Bayer CropScience developed "Liberty Link Corn" that is resistant to glufosinate. Pioneer Hi-Bred has developed and markets corn hybrids with tolerance to imidazoline herbicides under the trademark "Clearfield" - though in these hybrids, the herbicide-tolerance trait was bred using tissue culture selection and the chemical mutagen ethyl methanesulfonate, not genetic engineering. Consequently, the regulatory framework governing the approval of transgenic crops does not apply for Clearfield.

As of 2011, herbicide-resistant GM corn was grown in 14 countries. By 2012, 26 varieties herbicide-resistant GM maize were authorised for import into the European Union., but such imports remain controversial. Cultivation of herbicide-resistant corn in the EU provides substantial farm-level benefits.

Insecticide-producing Corn

The European corn borer, *Ostrinia nubilalis*, destroys corn crops by burrowing into the stem, causing the plant to fall over.

Bt corn is a variant of maize that has been genetically altered to express one or more proteins from the bacterium *Bacillus thuringiensis*. The protein is poisonous to certain insect pests and is widely used in organic gardening. The European corn borer causes about a billion dollars in damage to corn crops each year.

In recent years, traits have been added to ward off Corn ear worms and root worms, the latter of which annually causes about a billion dollars in damages.

The Bt protein is expressed throughout the plant. When a vulnerable insect eats the Bt-containing plant, the protein is activated in its gut, which is alkaline. In the alkaline environment the protein partially unfolds and is cut by other proteins, forming a toxin that paralyzes the insect's digestive system and forms holes in the gut wall. The insect stops eating within a few hours and eventually starves.

In 1996, the first GM maize producing a Bt Cry protein was approved, which killed the European corn borer and related species; subsequent Bt genes were introduced that killed corn rootworm larvae.

Approved Bt genes include single and stacked (event names bracketed) configurations of: Cry1A.105 (MON89034), CryIAb (MON810), CryIF (1507), Cry2Ab (MON89034), Cry3Bb1 (MON863 and

MON88017), Cry34Ab1 (59122), Cry35Ab1 (59122), mCry3A (MIR604), and Vip3A (MIR162), in both corn and cotton. Corn genetically modified to produce VIP was first approved in the US in 2010.

Drought Resistance

In 2013 Monsanto launched the first transgenic drought tolerance trait in a line of corn hybrids called DroughtGard. The MON 87460 trait is provided by the insertion of the cspB gene from the soil microbe *Bacillus subtilis*; it was approved by the USDA in 2011 and by China in 2013.

Sweet Corn

GM sweet corn varieties include "Attribute", the brand name for insect-resistant sweet corn developed by Syngenta. and Performance Series™ insect-resistant sweet corn developed by Monsanto.

Products in Development

In 2007, South African researchers announced the production of transgenic maize resistant to maize streak virus (MSV), although it has not been released as a product.

Refuges

US Environmental Protection Agency (EPA) regulations require farmers who plant Bt corn to plant non-Bt corn nearby (called a refuge) to provide a location to harbor vulnerable pests. Typically, 20% of corn in a grower's fields must be refuge; refuge must be at least 0.5 miles from Bt corn for lepidopteran pests, and refuge for corn rootworm must at least be adjacent to a Bt field.

The theory behind these refuges is to slow the evolution of resistance to the pesticide. EPA regulations also require seed companies to train farmers how to maintain refuges, to collect data on the refuges and to report that data to the EPA. A study of these reports found that from 2003 to 2005 farmer compliance with keeping refuges was above 90%, but that by 2008 approximately 25% of Bt corn farmers did not keep refuges properly, raising concerns that resistance would develop.

Unmodified crops received most of the economic benefits of Bt corn in the US in 1996-2007, because of the overall reduction of pest populations. This reduction came because females laid eggs on modified and unmodified strains alike.

Seed bags containing both Bt and refuge seed have been approved by the EPA in the United States. These seed mixtures were marketed as "Refuge in a Bag" (RIB) to increase farmer compliance with refuge requirements and reduce additional work needed at planting from having separate Bt and refuge seed bags on hand. The EPA approved a lower percentage of refuge seed in these seed mixtures ranging from 5 to 10%. This strategy is likely to reduce the likelihood of Bt-resistance occurring for corn rootworm, but may increase the risk of resistance for lepidopteran pests, such as European corn borer. Increased concerns for resistance with seed mixtures include partially resistant larvae on a Bt plant being able to move to a susceptible plant to survive or cross pollination of refuge pollen on to Bt plants that can lower the amount of Bt expressed in kernels for ear feeding insects.

Resistance

Resistant strains of the European corn borer have developed in areas with defective or absent refuge management.

In November 2009, Monsanto scientists found the pink bollworm had become resistant to first-generation Bt cotton in parts of Gujarat, India - that generation expresses one Bt gene, *Cry1Ac*. This was the first instance of Bt resistance confirmed by Monsanto anywhere in the world. Bollworm resistance to first generation Bt cotton has been identified in the Australia, China, Spain and the United States. In 2012, a Florida field trial demonstrated that army worms were resistant to pesticide-containing GM corn produced by Dupont-Dow; armyworm resistance was first discovered in Puerto Rico in 2006, prompting Dow and DuPont to voluntarily stop selling the product on the island.

Regulation

Regulation of GM crops varies between countries, with some of the most-marked differences occurring between the USA and Europe. Regulation varies in a given country depending on intended uses.

Controversy

There is a scientific consensus that currently available food derived from GM crops poses no greater risk to human health than conventional food, but that each GM food needs to be tested on a case-by-case basis before introduction. Nonetheless, members of the public are much less likely than scientists to perceive GM foods as safe. The legal and regulatory status of GM foods varies by country, with some nations banning or restricting them, and others permitting them with widely differing degrees of regulation.

The scientific rigor of the studies regarding human health has been disputed due to alleged lack of independence and due to conflicts of interest involving governing bodies and some of those who perform and evaluate the studies.

GM crops provide a number of ecological benefits, but there are also concerns for their overuse, stalled research outside of the Bt seed industry, proper management and issues with Bt resistance arising from their misuse.

Critics have objected to GM crops on ecological, economic and health grounds. The economic issues derive from those organisms that are subject to intellectual property law, mostly patents. The first generation of GM crops lose patent protection beginning in 2015. Monsanto has claimed it will not pursue farmers who retain seeds of off-patent varieties. These controversies have led to litigation, international trade disputes, protests and to restrictive legislation in most countries.

Effects on Nontarget Insects

Critics claim that Bt proteins could target predatory and other beneficial or harmless insects as well as the targeted pest. These proteins have been used as organic sprays for insect control in France since 1938 and the USA since 1958 with no ill effects on the environment reported. While cyt proteins are toxic towards the insect orders Coleoptera (beetles) and Diptera (flies), cry proteins selectively target Lepidopterans (moths and butterflies). As a toxic mechanism, cry proteins bind to specific receptors on the membranes of mid-gut (epithelial) cells, resulting in rupture of those cells. Any organism that lacks the appropriate gut receptors cannot be affected by the cry protein, and therefore Bt. Regulatory agencies assess the potential for the transgenic plant to impact nontarget organisms before approving commercial release.

A 1999 study found that in a lab environment, pollen from Bt maize dusted onto milkweed could harm the monarch butterfly. Several groups later studied the phenomenon in both the field and the laboratory, resulting in a risk assessment that concluded that any risk posed by the corn to butterfly populations under real-world conditions was negligible. A 2002 review of the scientific literature concluded that "the commercial large-scale cultivation of current Bt–maize hybrids did not pose a significant risk to the monarch population". A 2007 review found that "nontarget invertebrates are generally more abundant in Bt cotton and Bt maize fields than in nontransgenic fields managed with insecticides. However, in comparison with insecticide-free control fields, certain nontarget taxa are less abundant in Bt fields."

Gene Flow

Gene flow is the transfer of genes and/or alleles from one species to another. Concerns focus on the interaction between GM and other maize varieties in Mexico, and of gene flow into refuges.

In 2009 the government of Mexico created a regulatory pathway for genetically modified maize, but because Mexico is the center of diversity for maize, gene flow could affect a large fraction of the world's maize strains. A 2001 report in *Nature* presented evidence that Bt maize was cross-breeding with unmodified maize in Mexico. The data in this paper was later described as originating from an artifact. *Nature* later stated, "the evidence available is not sufficient to justify the publication of the original paper". A 2005 large-scale study failed to find any evidence of contamination in Oaxaca. However, other authors also found evidence of cross-breeding between natural maize and transgenic maize.

A 2004 study found Bt protein in kernels of refuge corn.

Food

The French High Council of Biotechnologies Scientific Committee reviewed the 2009 Vendômois et al. study and concluded that it "..presents no admissible scientific element likely to ascribe any haematological, hepatic or renal toxicity to the three re-analysed GMOs." However, the French government applies the precautionary principle with respect to GMOs.

A review by Food Standards Australia New Zealand and others of the same study concluded that the results were due to chance alone.

A 2011 Canadian study looked at the presence of CryAb1 protein (BT toxin) in non-pregnant women, pregnant women and fetal blood. All groups had detectable levels of the protein, including 93% of pregnant women and 80% of fetuses at concentrations of 0.19 ± 0.30 and 0.04 ± 0.04 mean ± SD ng/ml, respectively. The paper did not discuss safety implications or find any health problems. The paper was found to be unconvincing by multiple authors and organizations. In a swine model,

Cry1Ab-specific antibodies were not detected in pregnant sows or their offspring and no negative effects from feeding Bt maize to pregnant sows were observed.

In January 2013, the European Food Safety Authority released all data submitted by Monsanto in relation to the 2003 authorisation of maize genetically modified for glyphosate tolerance.

Starlink Corn Recalls

StarLink contains Cry9C, which had not previously been used in a GM crop. Starlink's creator, Plant Genetic Systems had applied to the US Environmental Protection Agency (EPA) to market Starlink for use in animal feed and in human food. However, because the Cry9C protein lasts longer in the digestive system than other Bt proteins, the EPA had concerns about its allergenicity, and PGS did not provide sufficient data to prove that Cry9C was not allergenic. As a result, PGS split its application into separate permits for use in food and use in animal feed. Starlink was approved by the EPA for use in animal feed only in May 1998.

StarLink corn was subsequently found in food destined for consumption by humans in the US, Japan, and South Korea. This corn became the subject of the widely publicized Starlink corn recall, which started when Taco Bell-branded taco shells sold in supermarkets were found to contain the corn. Sales of StarLink seed were discontinued. The registration for Starlink varieties was voluntarily withdrawn by Aventis in October 2000. (Pioneer had been bought by AgrEvo which then became Aventis CropScience at the time of the incident, which was later bought by Bayer

Fifty-one people reported adverse effects to the FDA; US Centers for Disease Control (CDC), which determined that 28 of them were possibly related to Starlink. However, the CDC studied the blood of these 28 individuals and concluded there was no evidence of hypersensitivity to the Starlink Bt protein.

A subsequent review of these tests by the Federal Insecticide, Fungicide, and Rodenticide Act Scientific Advisory Panel points out that while "the negative results decrease the probability that the Cry9C protein is the cause of allergic symptoms in the individuals examined ... in the absence of a positive control and questions regarding the sensitivity and specificity of the assay, it is not possible to assign a negative predictive value to this."

The US corn supply has been monitored for the presence of the Starlink Bt proteins since 2001.

In 2005, aid sent by the UN and the US to Central American nations also contained some StarLink corn. The nations involved, Nicaragua, Honduras, El Salvador and Guatemala refused to accept the aid.

Corporate Espionage

On December 19, 2013 six Chinese citizens were indicted in Iowa on charges of plotting to steal genetically modified seeds worth tens of millions of dollars from Monsanto and DuPont. Mo Hailong, director of international business at the Beijing Dabeinong Technology Group Co., part of the Beijing-based DBN Group, was accused of stealing trade secrets after he was found digging in an Iowa cornfield.

Genetically Modified Soybean

A genetically modified soybean is a soybean (*Glycine max*) that has had DNA introduced into it using genetic engineering techniques. In 1994 the first genetically modified soybean was introduced to the U.S. market, by Monsanto. In 2014, 90.7 million hectares of GM soy were planted worldwide, 82% of the total soy cultivation area.

Examples of Transgenic Soybeans

The genetic makeup of a soybean gives it a wide variety of uses, thus keeping it in high demand. First, manufacturers only wanted to use transgenics to be able to grow more soy at a minimal cost to meet this demand, and to fix any problems in the growing process, but they eventually found they could modify the soybean to contain healthier components, or even focus on one aspect of the soybean to produce in larger quantities. These phases became known as the first and second generation of genetically modified (GM) foods. As Peter Celec describes, "benefits of the first generation of GM foods were oriented towards the production process and companies, the second generation of GM foods offers, on contrary, various advantages and added value for the consumer", including "improved nutritional composition or even therapeutic effects."

Roundup Ready Soybean

Roundup Ready Soybeans (The first variety was also known as GTS 40-3-2 (OECD UI: MON-04032-6)) are a series of genetically engineered varieties of glyphosate-resistant soybeans produced by Monsanto.

Glyphosate kills plants by interfering with the synthesis of the essential amino acids phenylalanine, tyrosine and tryptophan. These amino acids are called "essential" because animals cannot make them; only plants and micro-organisms can make them and animals obtain them by eating plants.

Plants and microorganisms make these amino acids with an enzyme that only plants and lower organisms have, called 5-enolpyruvylshikimate-3-phosphate synthase (EPSPS). EPSPS is not present in animals, which instead obtain aromatic amino acids from their diet.

Roundup Ready Soybeans express a version of EPSPS from the CP4 strain of the bacteria, *Agrobacterium tumefaciens*, expression of which is regulated by an enhanced 35S promoter (E35S) from cauliflower mosaic virus (CaMV), a chloroplast transit peptide (CTP4) coding sequence from Petunia hybrida, and a nopaline synthase (nos 3') transcriptional termination element from Agrobacterium tumefaciens. The plasmid with EPSPS and the other genetic elements mentioned above was inserted into soybean germplasm with a gene gun by scientists at Monsanto and Asgrow. The patent on the first generation of Roundup Ready soybeans expired in March 2015.

History

First approved commercially in the United States during 1994, GTS 40-3-2 was subsequently introduced to Canada in 1995, Japan and Argentina in 1996, Uruguay in 1997, Mexico and Brazil in 1998, and South Africa in 2001.

Detection

GTS 40-3-2 can be detected using both nucleic acid and protein analysis methods.

Generic GMO Soybeans

Following expiration of Monsanto's patent on the first variety of glyphosate-resistant Roundup Ready soybeans, development began on glyphosate-resistant "generic" soybeans. The first variety, developed at the University of Arkansas Division of Agriculture, came on the market in 2015. With a slightly lower yield than newer Monsanto varieties, it costs about half as much, and seeds can be saved for subsequent years. According to its creator it is adapted to conditions in Arkansas. Several other varieties are being bred by crossing the original variety of Roundup Ready soybeans with other soybean varieties.

Stacked Traits

Monsanto developed a glyphosate-resistant soybean that also expresses Cry1Ac protein from Bacillus thuringiensis and the glyphosate-resistance gene, which completed the Brazilian regulatory process in 2010.

Genetic Modification to Improve Soybean oil

Soy has been genetically modified to improve the quality of soy oil. Soy oil has a fatty acid profile that makes it susceptible to oxidation, which makes it rancid, and this has limited its usefulness to the food industry. Genetic modifications increased the amount of oleic acid and stearic acid and decreased the amount of linolenic acid. By silencing, or knocking out, the delta 9 and delta 12 desaturases. DuPont Pioneer created a high oleic fatty acid soybean with levels of oleic acid greater than 80%, and started marketing it in 2010.

Regulation

The regulation of genetic engineering concerns the approaches taken by governments to assess and manage the risks associated with the development and release of genetically modified crops. There are differences in the regulation of GM crops between countries, with some of the most marked differences occurring between the USA and Europe. Regulation varies in a given country depending on the intended use of the products of the genetic engineering. For example, a crop not intended for food use is generally not reviewed by authorities responsible for food safety.

Controversy

There is a scientific consensus that currently available food derived from GM crops poses no greater risk to human health than conventional food, but that each GM food needs to be tested on a case-by-case basis before introduction. Nonetheless, members of the public are much less likely than scientists to perceive GM foods as safe. The legal and regulatory status of GM foods varies by country, with some nations banning or restricting them, and others permitting them with widely differing degrees of regulation.

A 2010 study found that in the United States, GM crops also provide a number of ecological benefits.

Critics have objected to GM crops on several grounds, including ecological concerns, and economic concerns raised by the fact these organisms are subject to intellectual property law. GM crops also are involved in controversies over GM food with respect to whether food produced from GM crops is safe and whether GM crops are needed to address the world's food needs.

Genetically Modified Tree

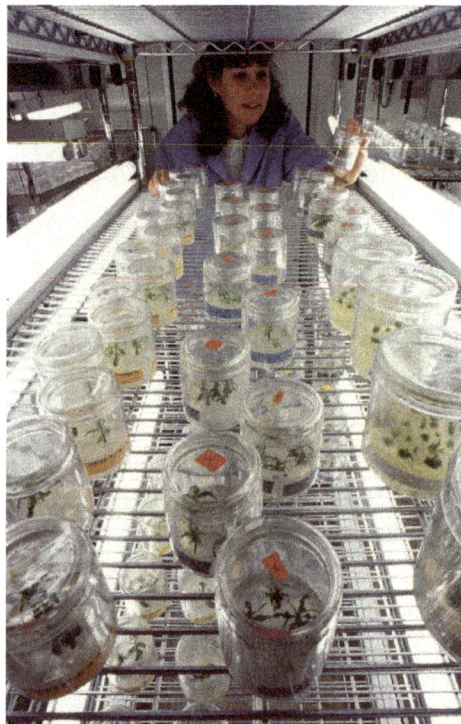

Technician checks on genetically modified peach and apple "orchards". Each dish holds experimental trees grown from lab-cultured cells to which researchers have given new genes. Source: USDA.

A genetically modified tree (GMt, GM tree, genetically engineered tree, GE tree or transgenic tree) is a tree whose DNA has been modified using genetic engineering techniques. In most cases the aim is to introduce a novel trait to the plant which does not occur naturally within the species. Examples include resistance to certain pests, diseases, environmental conditions, and herbicide tolerance, or the alteration of lignin levels in order to reduce pulping costs.

Genetically modified forest trees are not yet approved ("deregulated") for commercial use, with the exception of insect-resistant poplar trees in China. and one case of GM Eucalyptus in Brazil. Several genetically modified forest tree species are undergoing field trials for deregulation, and much of the research is being carried out by the pulp and paper industry, primarily with the intention of increasing the productivity of existing tree stock. Certain genetically modified orchard tree species have been deregulated for commercial use in the United States including the papaya and plum. The development, testing and use of GM trees remains at an early stage in comparison to GM crops.

Research

Research into genetically modified trees has been ongoing since 1988. Concerns surrounding the biosafety implications of releasing genetically modified trees into the wild have held back regulatory approval of GM forest trees. This concern is exemplified in the Convention on Biological Diversity's stance:

The Conference of the Parties, Recognising the uncertainties related to the potential environmental and socio-economic impacts, including long term and trans-boundary impacts, of genetically modified trees on global forest biological diversity, as well as on the livelihoods of indigenous and local communities, and given the absence of reliable data and of capacity in some countries to undertake risk assessments and to evaluate those potential impacts, recommends parties to take a precautionary approach when addressing the issue of genetically modified trees.

A precondition for further commercialization of GM forest trees is likely to be their complete sterility. Plantation trees remain phenotypically similar to their wild cousins in that most are the product of no more than three generations of artificial selection, therefore, the risk of transgene escape by pollination with compatible wild species is high. One of the most credible science based concerns with GM trees is their potential for wide dispersal of seed and pollen. The fact that pine pollen travels long distances is well established, moving up to 3,000 kilometers from its source. Additionally, many tree species reproduce for a long time before being harvested. In combination these factors have led some to believe that GM trees are worthy of special environmental considerations over GM crops. Ensuring sterility for GM trees has proven elusive, but efforts are being made. While tree geneticist Steve Strauss predicted that complete containment might be possible by 2020, many questions remain.

Proposed Uses

GM trees under experimental development have been modified with traits intended to provide benefit to industry, foresters or consumers. Due to high regulatory and research costs, the majority of genetically modified trees in silviculture consist of plantation trees, such as eucalyptus, poplar, and pine.

Lignin Alteration

Several companies in the pulp and paper industry are interested in utilizing GM technology to alter the lignin content of plantation trees. It is estimated that reducing lignin in plantation trees by genetic modification could reduce pulping costs by up to $15 per cubic metre. Lignin removal from wood fibres conventionally relies on costly and environmentally hazardous chemicals. By developing low-lignin GM trees it is hoped that pulping and bleaching processes will require fewer inputs, therefore, mills supplied by low-lignin GM trees may have a reduced impact on their surrounding ecosystems and communities. However, it is argued that reductions in lignin may compromise the structural integrity of the plant, thereby making it more susceptible to pathogens and disease, which could necessitate pesticide use exceeding that on traditional plantations.

Frost Tolerance

Genetic modification can allow trees to cope with abiotic stresses such that their geographic range is broadened. Freeze-tolerant GM eucalyptus trees for use in southern US plantations are currently being tested in open air sites with such an objective in mind. ArborGen, a tree biotechnology company and joint venture of pulp and paper firms Rubicon (New Zealand), MeadWestvaco (US) and International Paper (US) is leading this research. Until now the cultivation of eucalyptus has only been possible on the southern tip of Florida, freeze-tolerance would substantially extend the cultivation range northwards.

Accelerated Growth

In Brazil, field trials of fast growing GM eucalyptus are currently underway, they are set to conclude in 2015-2016 with commercialization to result. *FuturaGene*, a biotechnology company owned by Suzano, a Brazilian pulp and paper company, has been leading this research. Stanley Hirsch, chief executive of FuturaGene has stated: "Our trees grow faster and thicker. We are ahead of everyone. We have shown we can increase the yields and growth rates of trees more than anything grown by traditional breeding." The company is looking to reduce harvest cycles from 7 to 5.5 years with 20-30% more mass than conventional eucalyptus. There is concern that such objectives may further exacerbate the negative impacts of plantation forestry. Increased water and soil nutrient demand from faster growing species may lead to irrecoverable losses in site productivity and further impinge upon neighbouring communities and ecosystems.

Disease Resistance

Ecologically motivated research into genetic modification is underway. There are ongoing schemes that aim to foster disease resistance in trees such as the American chestnut and the English elm for the purpose of their reintroduction to the wild. Specific diseases have reduced the populations of these emblematic species to the extent that they are mostly lost in the wild. Genetic modification is being pursued concurrently with traditional breeding techniques in an attempt to endow these species with disease resistance.

Current Uses

Poplars in China

In 2002 China's State Forestry Administration approved GM poplar trees for commercial use. Subsequently 1.4 million Bt (insecticide) producing GM poplars were planted in China. They were planted both for their wood and as part of China's 'Green Wall' project, which aims to impede desertification. Reports indicate that the GM poplars have spread beyond the area of original planting and that contamination of native poplars with the Bt gene is occurring. There is concern with these developments, particularly because the pesticide producing trait may impart a positive selective advantage on the poplar, allowing it a high level of invasiveness.

References

- Martins VAP (2008). "Genomic Insights into Oil Biodegradation in Marine Systems". Microbial Biodegradation: Genomics and Molecular Biology. Caister Academic Press. ISBN 978-1-904455-17-2.

- Praitis, Vida (2006). "Creation of Transgenic Lines Using Microparticle Bombardment Methods". 351: 93–108. doi:10.1385/1-59745-151-7:93. ISBN 1-59745-151-7.

- Hayward, M.D.; Bosemark, N.O.; Romagosa, T. (2012). Plant Breeding: Principles and Prospects. Springer Science & Business Media. p. 131. ISBN 9789401115247.

- Staff (2009) Global Prevalence Of Vitamin A Deficiency in Populations At Risk 1995–2005 WHO Global Database on Vitamin A Deficiency. Geneva, World Health Organization, ISBN 978-92-4-159801-9, Retrieved 10 October 2011

- Thomson JA. "Genetic Engineering of Plants" (PDF). Biotechnology. Encyclopedia of Life Support Systems. 3. Retrieved 17 July 2016.

- Cotter, Jannet. "The Broken Promises of "Golden" Rice" (PDF). Greenpeace International. Greenpeace International. Retrieved 8 August 2016.

- Achenbach, Joel (2016-06-30). "107 Nobel laureates sign letter blasting Greenpeace over GMOs". Washington Post. Retrieved 2016-07-11.

- Dobson, Roger (2000), Royalty-free licenses for genetically modified rice made available to developing countries (PDF), 78 (10), Bulletin of the World Health Organisation, p. 1281, PMC 2560613, retrieved January 7, 2016 .

- Final Report of the PABE research project (December 2001). "Public Perceptions of Agricultural Biotechnologies in Europe". Commission of European Communities. Retrieved February 24, 2016.

- "Restrictions on Genetically Modified Organisms". Library of Congress. June 9, 2015. Retrieved February 24, 2016.

- Bashshur, Ramona (February 2013). "FDA and Regulation of GMOs". American Bar Association. Retrieved February 24, 2016.

- "Pocket K No. 16: Global Status of Commercialized Biotech/GM Crops in 2014". isaaa.org. International Service for the Acquisition of Agri-biotech Applications. Retrieved 23 February 2016.

- Final Report of the PABE research project (December 2001). "Public Perceptions of Agricultural Biotechnologies in Europe". Commission of European Communities. Retrieved February 24, 2016.

- Bashshur, Ramona (February 2013). "FDA and Regulation of GMOs". American Bar Association. Retrieved February 24, 2016.

- Lynch, Diahanna; Vogel, David (April 5, 2001). "The Regulation of GMOs in Europe and the United States: A Case-Study of Contemporary European Regulatory Politics". Council on Foreign Relations. Retrieved February 24, 2016.

Plant Breeding: An Overview

Plant breeding, like animal breeding, is introduced in plants to produce favorable characteristics in plants. A cultigen is a plant that is purposely selected by humans and happens because of artificial selection. The following text on plant breeding is an overview of the subject matter, incorporating all the major aspects of plant breeding.

Plant Breeding

The Yecoro wheat (right) cultivar is sensitive to salinity, plants resulting from a hybrid cross with cultivar W4910 (left) show greater tolerance to high salinity

Plant breeding is the art and science of changing the traits of plants in order to produce desired characteristics. Plant breeding can be accomplished through many different techniques ranging from simply selecting plants with desirable characteristics for propagation, to more complex molecular techniques.

Plant breeding has been practiced for thousands of years, since near the beginning of human civilization. It is practiced worldwide by individuals such as gardeners and farmers, or by professional plant breeders employed by organizations such as government institutions, universities, crop-specific industry associations or research centers.

International development nation agencies believe that breeding new crops is important for ensuring food security by developing new varieties that are higher-yielding, disease resistant, drought-resistant or regionally adapted to different environments and growing conditions.

History

breeding started with sedentary agriculture and particularly the domestication of the first agricultural plants, a practice which is estimated to date back 9,000 to 11,000 years. Initially early farmers simply selected food plants with particular desirable characteristics, and employed these as progenitors for subsequent generations, resulting in an accumulation of valuable traits over time.

Gregor Mendel's experiments with plant hybridization led to his establishing laws of inheritance. Once this work became well known, it formed the basis of the new science of genetics, which stimulated research by many plant scientists dedicated to improving crop production through plant breeding.

Modern plant breeding is applied genetics, but its scientific basis is broader, covering molecular biology, cytology, systematics, physiology, pathology, entomology, chemistry, and statistics (biometrics). It has also developed its own technology.

Classical Plant Breeding

One major technique of plant breeding is selection, the process of selectively propagating plants with desirable characteristics and eliminating or "culling" those with less desirable characteristics.

Another technique is the deliberate interbreeding (crossing) of closely or distantly related individuals to produce new crop varieties or lines with desirable properties. Plants are crossbred to introduce traits/genes from one variety or line into a new genetic background. For example, a mildew-resistant pea may be crossed with a high-yielding but susceptible pea, the goal of the cross being to introduce mildew resistance without losing the high-yield characteristics. Progeny from the cross would then be crossed with the high-yielding parent to ensure that the progeny were most like the high-yielding parent, (backcrossing). The progeny from that cross would then be tested for yield (selection, as described above) and mildew resistance and high-yielding resistant plants would be further developed. Plants may also be crossed with themselves to produce inbred varieties for breeding. Pollinators may be excluded through the use of pollination bags.

Classical breeding relies largely on homologous recombination between chromosomes to generate genetic diversity. The classical plant breeder may also make use of a number of *in vitro* techniques such as protoplast fusion, embryo rescue or mutagenesis to generate diversity and produce hybrid plants that would not exist in nature.

Traits that breeders have tried to incorporate into crop plants include:

- Improved quality, such as increased nutrition, improved flavor, or greater beauty
- Increased yield of the crop
- Increased tolerance of environmental pressures (salinity, extreme temperature, drought)
- Resistance to viruses, fungi and bacteria
- Increased tolerance to insect pests
- Increased tolerance of herbicides
- Longer storage period for the harvested crop

Before World War Ii

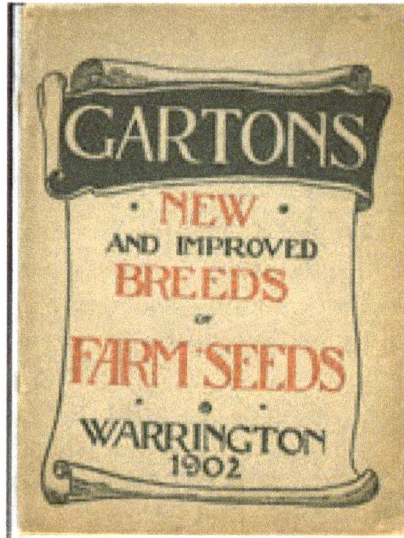

Garton's catalogue from 1902

Successful commercial plant breeding concerns were founded from the late 19th century. Gartons Agricultural Plant Breeders in England was established in the 1890s by John Garton, who was one of the first to commercialize new varieties of agricultural crops created through cross-pollination. The firm's first introduction was Abundance Oat, one of the first agricultural grain varieties bred from a *controlled* cross, introduced to commerce in 1892.

In the early 20th century, plant breeders realized that Mendel's findings on the non-random nature of inheritance could be applied to seedling populations produced through deliberate pollinations to predict the frequencies of different types. Wheat hybrids were bred to increase the crop production of Italy during the so-called "Battle for Grain" (1925–1940). Heterosis was explained by George Harrison Shull. It describes the tendency of the progeny of a specific cross to outperform both parents. The detection of the usefulness of heterosis for plant breeding has led to the development of inbred lines that reveal a heterotic yield advantage when they are crossed. Maize was the first species where heterosis was widely used to produce hybrids.

Statistical methods were also developed to analyze gene action and distinguish heritable variation from variation caused by environment. In 1933 another important breeding technique, cytoplasmic male sterility (CMS), developed in maize, was described by Marcus Morton Rhoades. CMS is a maternally inherited trait that makes the plant produce sterile pollen. This enables the production of hybrids without the need for labor-intensive detasseling.

These early breeding techniques resulted in large yield increase in the United States in the early 20th century. Similar yield increases were not produced elsewhere until after World War II, the Green Revolution increased crop production in the developing world in the 1960s.

After World War Ii

Following World War II a number of techniques were developed that allowed plant breeders to hybridize distantly related species, and artificially induce genetic diversity.

When distantly related species are crossed, plant breeders make use of a number of plant tissue culture techniques to produce progeny from otherwise fruitless mating. Interspecific and intergeneric hybrids are produced from a cross of related species or genera that do not normally sexually reproduce with each other. These crosses are referred to as *Wide crosses*. For example, the cereal triticale is a wheat and rye hybrid. The cells in the plants derived from the first generation created from the cross contained an uneven number of chromosomes and as result was sterile. The cell division inhibitor colchicine was used to double the number of chromosomes in the cell and thus allow the production of a fertile line.

In vitro-culture of Vitis (grapevine), Geisenheim Grape Breeding Institute

Failure to produce a hybrid may be due to pre- or post-fertilization incompatibility. If fertilization is possible between two species or genera, the hybrid embryo may abort before maturation. If this does occur the embryo resulting from an interspecific or intergeneric cross can sometimes be rescued and cultured to produce a whole plant. Such a method is referred to as Embryo Rescue. This technique has been used to produce new rice for Africa, an interspecific cross of Asian rice *(Oryza sativa)* and African rice *(Oryza glaberrima)*.

Hybrids may also be produced by a technique called protoplast fusion. In this case protoplasts are fused, usually in an electric field. Viable recombinants can be regenerated in culture.

Chemical mutagens like EMS and DMS, radiation and transposons are used to generate mutants with desirable traits to be bred with other cultivars - a process known as *Mutation Breeding*. Classical plant breeders also generate genetic diversity within a species by exploiting a process called somaclonal variation, which occurs in plants produced from tissue culture, particularly plants derived from callus. Induced polyploidy, and the addition or removal of chromosomes using a technique called chromosome engineering may also be used.

When a desirable trait has been bred into a species, a number of crosses to the favored parent are made to make the new plant as similar to the favored parent as possible. Returning to the example of the mildew resistant pea being crossed with a high-yielding but susceptible pea, to make the mildew resistant progeny of the cross most like the high-yielding parent, the progeny will be crossed back to that parent for several generations. This process removes most of the genetic contribution of the mildew resistant parent. Classical breeding is therefore a cyclical process.

With classical breeding techniques, the breeder does not know exactly what genes have been introduced to the new cultivars. Some scientists therefore argue that plants produced by classical breeding methods should undergo the same safety testing regime as genetically modified plants. There have been instances where plants bred using classical techniques have been unsuitable for human consumption, for example the poison solanine was unintentionally increased to unacceptable levels in certain varieties of potato through plant breeding. New potato varieties are often screened for solanine levels before reaching the marketplace.

Modern Plant Breeding

Modern plant breeding may use techniques of molecular biology to select, or in the case of genetic modification, to insert, desirable traits into plants. Application of biotechnology or molecular biology is also known as molecular breeding.

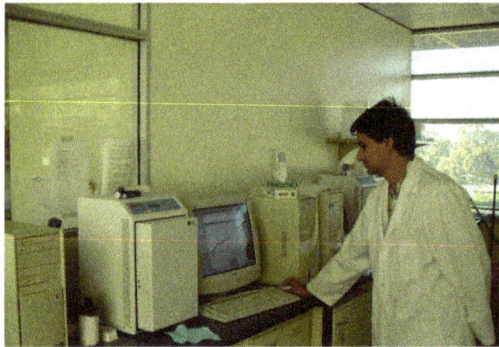

Modern facilities in molecular biology have converted classical plant breeding to molecular plant breeding

Steps of Plant Breeding

The following are the major activities of plant breeding:

1. Collection of variation
2. Selection
3. Evaluation
4. Release
5. Multiplication
6. Distribution of the new variety
7. selling to people

Marker Assisted Selection

Sometimes many different genes can influence a desirable trait in plant breeding. The use of tools such as molecular markers or DNA fingerprinting can map thousands of genes. This allows plant breeders to screen large populations of plants for those that possess the trait of interest. The screening is based on the presence or absence of a certain gene as determined by laboratory procedures, rather than on the visual identification of the expressed trait in the plant.

Reverse Breeding and Doubled Haploids (DH)

A method for efficiently producing homozygous plants from a heterozygous starting plant, which has all desirable traits. This starting plant is induced to produce doubled haploid from haploid cells, and later on creating homozygous/doubled haploid plants from those cells. While in natural offspring genetic recombination occurs and traits can be unlinked from each other, in doubled haploid cells and in the resulting DH plants recombination is no longer an issue. There, a recombination between two corresponding chromosomes does not lead to un-linkage of alleles or traits, since it just leads to recombination with its identical copy. Thus, traits on one chromosome stay linked. Selecting those offspring having the desired set of chromosomes and crossing them will result in a final F1 hybrid plant, having exactly the same set of chromosomes, genes and traits as the starting hybrid plant. The homozygous parental lines can reconstitute the original heterozygous plant by crossing, if desired even in a large quantity. An individual heterozygous plant can be converted into a heterozygous variety (F1 hybrid) without the necessity of vegetative propagation but as the result of the cross of two homozygous/doubled haploid lines derived from the originally selected plant. patent

Genetic Modification

Genetic modification of plants is achieved by adding a specific gene or genes to a plant, or by knocking down a gene with RNAi, to produce a desirable phenotype. The plants resulting from adding a gene are often referred to as transgenic plants. If for genetic modification genes of the species or of a crossable plant are used under control of their native promoter, then they are called cisgenic plants. Sometimes genetic modification can produce a plant with the desired trait or traits faster than classical breeding because the majority of the plant's genome is not altered.

To genetically modify a plant, a genetic construct must be designed so that the gene to be added or removed will be expressed by the plant. To do this, a promoter to drive transcription and a termination sequence to stop transcription of the new gene, and the gene or genes of interest must be introduced to the plant. A marker for the selection of transformed plants is also included. In the laboratory, antibiotic resistance is a commonly used marker: Plants that have been successfully transformed will grow on media containing antibiotics; plants that have not been transformed will die. In some instances markers for selection are removed by backcrossing with the parent plant prior to commercial release.

The construct can be inserted in the plant genome by genetic recombination using the bacteria *Agrobacterium tumefaciens* or *A. rhizogenes*, or by direct methods like the gene gun or microinjection. Using plant viruses to insert genetic constructs into plants is also a possibility, but the technique is limited by the host range of the virus. For example, Cauliflower mosaic virus (CaMV) only infects cauliflower and related species. Another limitation of viral vectors is that the virus is not usually passed on the progeny, so every plant has to be inoculated.

The majority of commercially released transgenic plants are currently limited to plants that have introduced resistance to insect pests and herbicides. Insect resistance is achieved through incorporation of a gene from *Bacillus thuringiensis* (Bt) that encodes a protein that is toxic to some insects. For example, the cotton bollworm, a common cotton pest, feeds on Bt cotton it will ingest the toxin and die. Herbicides usually work by binding to certain plant enzymes and inhibiting their action. The enzymes that the herbicide inhibits are known as the herbicides *target site*. Herbicide resistance can be engineered into crops by expressing a version of *target site* protein that is not inhibited by the herbicide. This is the method used to produce glyphosate resistant crop plants.

Genetic modification of plants that can produce pharmaceuticals (and industrial chemicals), sometimes called *pharming*, is a rather radical new area of plant breeding.

Issues and Concerns

Modern plant breeding, whether classical or through genetic engineering, comes with issues of concern, particularly with regard to food crops. The question of whether breeding can have a negative effect on nutritional value is central in this respect. Although relatively little direct research in this area has been done, there are scientific indications that, by favoring certain aspects of a plant's development, other aspects may be retarded. A study published in the *Journal of the American College of Nutrition* in 2004, entitled *Changes in USDA Food Composition Data for 43 Garden Crops, 1950 to 1999*, compared nutritional analysis of vegetables done in 1950 and in 1999, and found substantial decreases in six of 13 nutrients measured, including 6% of protein and 38% of riboflavin. Reductions in calcium, phosphorus, iron and ascorbic acid were also found. The study, conducted at the Biochemical Institute, University of Texas at Austin, concluded in summary: *"We suggest that any real declines are generally most easily explained by changes in cultivated varieties between 1950 and 1999, in which there may be trade-offs between yield and nutrient content."*

The debate surrounding genetically modified food during the 1990s peaked in 1999 in terms of media coverage and risk perception, and continues today - for example, *"Germany has thrown its weight behind a growing European mutiny over genetically modified crops by banning the planting of a widely grown pest-resistant corn variety."* The debate encompasses the ecological impact of genetically modified plants, the safety of genetically modified food and concepts used for safety evaluation like substantial equivalence. Such concerns are not new to plant breeding. Most countries have regulatory processes in place to help ensure that new crop varieties entering the marketplace are both safe and meet farmers' needs. Examples include variety registration, seed schemes, regulatory authorizations for GM plants, etc.

Plant breeders' rights is also a major and controversial issue. Today, production of new varieties is dominated by commercial plant breeders, who seek to protect their work and collect royalties through national and international agreements based in intellectual property rights. The range of related issues is complex. In the simplest terms, critics of the increasingly restrictive regulations argue that, through a combination of technical and economic pressures, commercial breeders are reducing biodiversity and significantly constraining individuals (such as farmers) from developing and trading seed on a regional level. Efforts to strengthen breeders' rights, for example, by lengthening periods of variety protection, are ongoing.

When new plant breeds or cultivars are bred, they must be maintained and propagated. Some plants are propagated by asexual means while others are propagated by seeds. Seed propagated cultivars require specific control over seed source and production procedures to maintain the integrity of the plant breeds results. Isolation is necessary to prevent cross contamination with related plants or the mixing of seeds after harvesting. Isolation is normally accomplished by planting distance but in certain crops, plants are enclosed in greenhouses or cages (most commonly used when producing F1 hybrids.)

Role of Plant Breeding in Organic Agriculture

Critics of organic agriculture claim it is too low-yielding to be a viable alternative to conventional agriculture. However, part of that poor performance may be the result of growing poorly adapted varieties. It is estimated that over 95% of organic agriculture is based on conventionally adapted varieties, even though the production environments found in organic vs. conventional farming systems are vastly different due to their distinctive management practices. Most notably, organic farmers have fewer inputs available than conventional growers to control their production environments. Breeding varieties specifically adapted to the unique conditions of organic agriculture is critical for this sector to realize its full potential. This requires selection for traits such as:

- Water use Efficiency
- Nutrient use efficiency (particularly nitrogen and phosphorus)
- Weed competitiveness
- Tolerance of mechanical weed control
- Pest/disease resistance
- Early maturity (as a mechanism for avoidance of particular stresses)
- Abiotic stress tolerance (i.e. drought, salinity, etc...)

Currently, few breeding programs are directed at organic agriculture and until recently those that did address this sector have generally relied on indirect selection (i.e. selection in conventional environments for traits considered important for organic agriculture). However, because the difference between organic and conventional environments is large, a given genotype may perform very differently in each environment due to an interaction between genes and the environment. If this interaction is severe enough, an important trait required for the organic environment may not be revealed in the conventional environment, which can result in the selection of poorly adapted individuals. To ensure the most adapted varieties are identified, advocates of organic breeding now promote the use of direct selection (i.e. selection in the target environment) for many agronomic traits.

There are many classical and modern breeding techniques that can be utilized for crop improvement in organic agriculture despite the ban on genetically modified organisms. For instance, controlled crosses between individuals allow desirable genetic variation to be recombined and transferred to seed progeny via natural processes. Marker assisted selection can also be employed as a diagnostics tool to facilitate selection of progeny who possess the desired trait(s), greatly speeding up the breeding process. This technique has proven particularly useful for the introgression of resistance genes into new backgrounds, as well as the efficient selection of many resistance genes pyramided into a single individual. Unfortunately, molecular markers are not currently available for many important traits, especially complex ones controlled by many genes.

Addressing Global Food Security Through Plant Breeding

For future agriculture to thrive there are necessary changes which must be made in accordance to arising global issues. These issues are arable land, harsh cropping conditions and food security which involves, being able to provide the world population with food containing sufficient nutrients. These crops need to be able to mature in several environments allowing for worldwide access, this is involves issues such as drought tolerance. These global issues are achievable through the process of plant breeding, as it offers the ability to select specific genes allowing the crop to perform at a level which yields the desired results.

Minimal Land Degradation

Land degradation is a major issue, as it can negatively impact the capability of the land to be productive. Poor agricultural management has a huge impact on the degradation of soil worldwide and it is Africa and Asia that are most affected. Through education and development of modified plants, these statistics can be reduced and agricultural land can become more productive. Plant breeding allows for an increase in yield with out the extra strain on the land. The genetically modified, Bt white maize, was introduced to South Africa and was surveyed in 33 large commercial farms and 368 small landholders properties and in both cases a higher yield was recorded.

Increased Yield Without Expansion

With an increasing population, the production of food needs to increase with it. It is estimated that a 70% increase in food production is needed by 2050 in order to meet the Declaration of the World Summit on Food Security. But with the natural degradation of agricultural land, simply planting more crops is no longer a viable option. Therefore, new varieties of plants need to be developed through plant breeding that generates an increase of yield without relying on an increase in land area. An example of this can be seen in Asia, where food production per capita has increased twofold. This has been achieved through not only the use of fertilisers, but through the use of better crops that have been specifically designed for the area.

Breeding for Increased Nutritional Value

Plant breeding can contribute to global food security as it is a cost-effective tool for increasing nutritional value of forage and crops. Improvements in nutritional value for forage crops from the use of analytical chemistry and rumen fermentation technology have been recorded since 1960; this science and technology gave breeders the ability to screen thousands of samples within a small amount of time, meaning breeders could identify a high performing hybrid quicker. The main area genetic increases were made was in vitro dry matter digestibility (IVDMD) resulting in 0.7-2.5% increase, at just 1% increase in IVDMD a single Bos Taurus also known as beef cattle reported 3.2% increase in daily gains. This improvement indicates plant breeding is an essential tool in gearing future agriculture to perform at a more advanced level.

Breeding for Tolerance

Plant breeding of hybrid crops has become extremely popular worldwide in an effort to combat the harsh environment. With long periods of drought and lack of water or nitrogen stress tolerance has become a significant part of agriculture. Plant breeders have focused on identifying crops which will ensure crops perform under these conditions; a way to achieve this is finding strains of the crop that is resistance to drought conditions with low nitrogen. It is evident from this that plant breeding is vital for future agriculture to survive as it enables farmers to produce stress resistant crops hence improving food security.

Participatory Plant Breeding

The development of agricultural science, with phenomenon like the Green Revolution arising, have left millions of farmers in developing countries, most of whom operate small farms under unstable and difficult growing conditions, in a precarious situation. The adoption of new plant varieties by this group has been hampered by the constraints of poverty and the international policies promoting an industrialized model of agriculture. Their response has been the creation of a novel and promising set of research methods collectively known as participatory plant breeding. Participatory means that farmers are more involved in the breeding process and breeding goals are defined by farmers instead of international seed companies with their large-scale breeding programs. Farmers' groups and NGOs, for example, may wish to affirm local people's rights over genetic resources, produce seeds themselves, build farmers' technical expertise, or develop new products for niche markets, like organically grown food.

List of Notable Plant Breeders

- Gartons Agricultural Plant Breeders
- Gregor Mendel
- Keith Downey
- Luther Burbank
- Nazareno Strampelli
- Niels Ebbesen Hansen
- Edger McFadden
- Norman Borlaug

Marker-assisted Selection

Marker assisted selection or marker aided selection (MAS) is an indirect selection process where a trait of interest is selected based on a marker (morphological, biochemical or DNA/RNA variation) linked to a trait of interest (e.g. productivity, disease resistance, abiotic stress tolerance, and quality), rather than on the trait itself. This process is used in plant and animal breeding.

For example, using MAS to select individuals with disease resistance involves identifying a marker allele that is linked with disease resistance rather than the level of disease resistance. The assumption is that the marker associates at high frequency with the gene or quantitative trait locus (QTL) of interest, due to genetic linkage (close proximity, on the chromosome, of the marker locus and the disease resistance-determining locus). MAS can be useful to select for traits that are difficult or expensive to measure, exhibit low heritability and/or are expressed late in development. At certain points in the breeding process the specimens are examined to ensure that they express the desired trait.

Marker Types

The majority of MAS work in the present era uses DNA-based markers. However, the first markers that allowed indirect selection of a trait of interest were morphological markers. In 1923, Sax[who?] first reported association of a simply inherited genetic marker with a quantitative trait in plants when he observed segregation of seed size associated with segregation for a seed coat color marker in beans (*Phaseolus vulgaris* L.). In 1935, Rasmusson demonstrated linkage of flowering time (a quantitative trait) in peas with a simply inherited gene for flower color.

Markers may be:

- Morphological - These markers are often detectable by eye, by simple visual inspection. Examples of this type of marker include the presence or absence of an awn, leaf sheath coloration, height, grain color, aroma of rice etc. In well-characterized crops like maize, tomato, pea, barley or wheat, tens or hundreds of genes that determine morphological traits have been mapped to specific chromosome locations.

- Biochemical- A protein that can be extracted and observed; for example, isozymes and storage proteins.

- Cytological - The chromosomal banding produced by different stains; for example, G banding.

- DNA-based or molecular- A unique gene (DNA sequence), occurring in proximity to the gene or locus of interest, can be identified by a range of molecular techniques such as RFLP, RAPD, AFLP, DAF, SCAR, microsatellite, or single-nucleotide polymorphism (SNP) detection.

Positive and Negative Selectable Markers

The following terms are generally less relevant to discussions of MAS in plant and animal breeding, but are highly relevant in molecular biology research:

- Positive selectable markers are selectable markers that confer selective advantage to the host organism. An example would be antibiotic resistance, which allows the host organism to survive antibiotic selection.

- Negative selectable markers are selectable markers that eliminate or inhibit growth of the host organism upon selection. An example would be thymidine kinase, which makes the host sensitive to ganciclovir selection.

A distinction can be made between selectable markers (which eliminate certain genotypes from the population) and screenable markers (which cause certain genotypes to be readily identifiable, at which point the experimenter must "score" or evaluate the population and act to retain the preferred genotypes). Most MAS uses screenable markers rather than selectable markers.

Gene Vs Marker

The gene of interest directly causes production of protein(s) or RNA that produce a desired trait or phenotype, whereas markers (a DNA sequence or the morphological or biochemical markers produced due to that DNA) are genetically linked to the gene of interest. The gene of interest and the marker tend to move together during segregation of gametes due to their proximity on the same chromosome and concomitant reduction in recombination (chromosome crossover events) between the marker and gene of interest. For some traits, the gene of interest has been discovered and the presence of desirable alleles can be directly assayed with a high level of confidence. However, if the gene of interest is not known, markers linked to the gene of interest can still be used to select for individuals with desirable alleles of the gene of interest. When markers are used there may be some inaccurate results due to inaccurate tests for the marker. There also can be false positive results when markers are used, due to recombination between the marker of interest and gene (or QTL). A perfect marker would elicit no false positive results. The term 'perfect marker' is sometimes used when tests are performed to detect a SNP or other DNA polymorphism in the gene of interest, if that SNP or other polymorphism is the direct cause of the trait of interest. The term 'marker' is still appropriate to use when directly assaying the gene of interest, because the test of genotype is an indirect test of the trait or phenotype of interest.

Important Properties of Ideal Markers for MAS

An ideal marker:

- Easy recognition of all possible phenotypes (homo- and heterozygotes) from all different alleles

- Demonstrates measurable differences in expression between trait types or gene of interest alleles, early in the development of the organism

- Testing for the marker does not have variable success depending on the allele at the marker locus or the allele at the target locus (the gene of interest that determines the trait of interest).

- Low or null interaction among the markers allowing the use of many at the same time in a segregating population

- Abundant in number

- Polymorphic

Demerits of Morphological Markers

Morphological markers are associated with several general deficits that reduce their usefulness including:

- the delay of marker expression until late into the development of the organism

- dominance

- deleterious effects

- pleiotropy

- confounding effects of genes unrelated to the gene or trait of interest but which also affect the morphological marker (epistasis)

- rare polymorphism

- frequent confounding effects of environmental factors which affect the morphological characteristics of the organism

To avoid problems specific to morphological markers, the DNA-based markers have been developed. They are highly polymorphic, exhibit simple inheritance (often codomimant), are abundant throughout the genome, are easy and fast to detect, exhibit minimum pleiotropic effects, and detection is not dependent on the developmental stage of the organism. Numerous markers have been mapped to different chromosomes in several crops including rice, wheat, maize, soybean and several others, and in livestock such as cattle, pigs and chickens. Those markers have been used in diversity analysis, parentage detection, DNA fingerprinting, and prediction of hybrid performance. Molecular markers are useful in indirect selection processes, enabling manual selection of individuals for further propagation.

Selection for Major Genes Linked to Markers

'Major genes' that are responsible for economically important characteristics are frequent in the plant kingdom. Such characteristics include disease resistance, male sterility, self-incompatibility, and others related to shape, color, and architecture of whole plants and are often of mono- or oligogenic in nature. The marker loci that are tightly linked to major genes can be used for selection and are sometimes more efficient than direct selection for the target gene. Such advantages in efficiency may be due for example, to higher expression of the marker mRNA in such cases that the marker is actually a gene. Alternatively, in such cases that the target gene of interest differs between two alleles by a difficult-to-detect single nucleotide polymorphism, an external marker (be it another gene or a polymorphism that is easier to detect, such as a short tandem repeat) may present as the most realistic option.

Situations that are Favorable for Molecular Marker Selection

There are several indications for the use of molecular markers in the selection of a genetic trait.

In such situations that:

- the selected character is expressed late in plant development, like fruit and flower features or adult characters with a juvenile period (so that it is not necessary to wait for the organism to become fully developed before arrangements can be made for propagation)

- the expression of the target gene is recessive (so that individuals which are heterozygous positive for the recessive allele can be crossed to produce some homozygous offspring with the desired trait)

- there is requirement for the presence of special conditions in order to invoke expression of the target gene(s), as in the case of breeding for disease and pest resistance (where inoculation with the disease or subjection to pests would otherwise be required). This advantage derives from the errors due to unreliable inoculation methods and the fact that field inoculation with the pathogen is not allowed in many areas for safety reasons. Moreover, problems in the recognition of the environmentally unstable genes can be eluded.

- the phenotype is affected by two or more unlinked genes (epistatis). For example, selection for multiple genes which provide resistance against diseases or insect pests for gene pyramiding.

The cost of genotyping (an example of a molecular marker assay) is reducing while the cost of phenotyping is increasing particularly in developed countries thus increasing the attractiveness of MAS as the development of the technology continues.

Steps for MAS

Generally the first step is to map the gene or quantitative trait locus (QTL) of interest first by using different techniques and then using this information for marker assisted selection. Generally, the markers to be used should be close to gene of interest (<5 recombination unit or cM) in order to ensure that only minor fraction of the selected individuals will be recombinants. Generally, not only a single marker but rather two markers are used in order to reduce the chances of an error due to homologous recombination. For example, if two flanking markers are used at same time with an interval between them of approximately 20cM, there is higher probability (99%) for recovery of the target gene.

QTL Mapping Techniques

In plants QTL mapping is generally achieved using bi-parental cross populations; a cross between two parents which have a contrasting phenotype for the trait of interest are developed. Commonly used populations are recombinant inbred lines (RILs), doubled haploids (DH), back cross and F_2. Linkage between the phenotype and markers which have already been mapped is tested in these populations in order to determine the position of the QTL. Such techniques are based on linkage and are therefore referred to as "linkage mapping".A

Single Step MAS and QTL Mapping

In contrast to two-step QTL mapping and MAS, a single-step method for breeding typical plant populations has been developed.

In such an approach, in the first few breeding cycles, markers linked to the trait of interest are identified by QTL mapping and later the same information is used in the same population. In this approach, pedigree structure is created from families that are created by crossing number of parents (in three-way or four way crosses). Both phenotyping and genotyping is done using molecular markers mapped the possible location of QTL of interest. This will identify markers and their favorable alleles. Once these favorable marker alleles are identified, the frequency of such alleles will be increased and response to marker assisted selection is estimated. Marker allele(s) with desirable effect will be further used in next selection cycle or other experiments.

High-throughput Genotyping Techniques

Recently high-throughput genotyping techniques are developed which allows marker aided screening of many genotypes. This will help breeders in shifting traditional breeding to marker aided selection. One example of such automation is using DNA isolation robots, capillary electrophoresis and pipetting robots.

One recent example of capllilary system is Applied Biosystems 3130 Genetic Analyzer. This is the latest generation of 4-capillary electrophoresis instruments for the low to medium throughput laboratories.

Use of MAS for Backcross Breeding

A minimum of five or six-backcross generations are required to transfer a gene of interest from a donor (may not be adapted) to a recipient (recurrent – adapted cultivar). The recovery of the recurrent genotype can be accelerated with the use of molecular markers. If the F1 is heterozygous for the marker locus, individuals with the recurrent parent allele(s) at the marker locus in first or subsequent backcross generations will also carry a chromosome tagged by the marker.

Marker Assisted Gene Pyramiding

Gene pyramiding has been proposed and applied to enhance resistance to disease and insects by selecting for two or more than two genes at a time. For example in rice such pyramids have been developed against bacterial blight and blast. The advantage of use of markers in this case allows to select for QTL-allele-linked markers that have same phenotypic effect.

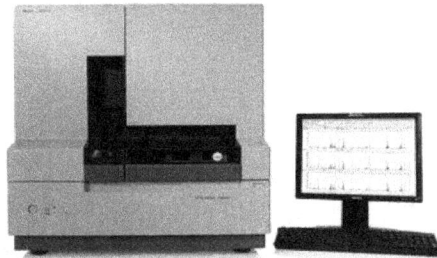

ABI 3130 genetic analyzer.

MAS has also been proved useful for livestock improvement.

A coordinated effort to implement wheat (*Triticum turgidum* and *Triticum aestivum*) marker assisted selection in the U.S. as well as a resource for marker assisted selection exists at the Wheat CAP (Coordinated Agricultural Project) website. Farhad Kahani

Marker Assisted Selection Learning Lessons

- Using OptiMAS for Marker-Assisted Recurrent Selection in a Multi-Parental Population
- Video Overview of DNA Extraction and Marker-Assisted Selection
- Marker Assisted Selection

- Background Selection

- Forward Selection

- Marker Assisted Selection in Tomato

- Marker-Assisted Selection for Bacterial Spot Resistance in Tomato Case Study

- Marker-Assisted Selection for PVY Resistance in Potato

- Marker-Assisted Selection for Resistance to Golden Nematode in Potato

- Gene Pyramiding Using Molecular Markers

- Pyramiding Resistance Genes for Bacterial Spot and Bacterial Speck on Tomato Chromosome 5

- The SolCAP Tomato Phenotypic Data: Estimating Heritability and BLUPs for Traits

- rrBLUP Package in R for Genomewide Selection

Cultigen

A cultigen is a plant that has been deliberately altered or selected by humans; it is the result of artificial selection. These "man-made" or anthro-pogenic plants are, for the most part, plants of commerce that are used in horticulture, agriculture and forestry. Because cultigens are defined by their mode of origin and not by where they are growing, plants meeting this definition remain cultigens whether they are naturalised in the wild, deliberately planted in the wild, or growing in cultivation. Cultigens arise in the following ways: selections of variants from the wild or cultivation including vegetative sports (aberrant growth that can be reproduced reliably in cultivation); plants that are the result of plant breeding and selection programs; genetically modified plants (plants modified by the deliberate implantation of genetic material); and graft-chimaeras (plants grafted to produce mixed tissue, the graft material possibly from wild plants, special selections, or hybrids).

Cultigens may be named in any of a number of ways. The traditional method of scientific naming is under the *International Code of Nomenclature for algae, fungi, and plants*, and many of the most important cultigens, like maize (*Zea mays*) and banana (*Musa acuminata*), are so named. Although it is perfectly in order to give a cultigen a botanical name, in any rank desired, now or at any other time, these days it is more common for cultigens to be given names in accordance with the principles, rules and recommendations laid down in the International Code of Nomenclature for Cultivated Plants (ICNCP) which provides for the names of cultigens in three classification categories, the cultivar, the Group (formerly cultivar-group), and the grex. From that viewpoint it may be said that there is a separate discipline of cultivated plant taxonomy, which forms one of the ways to look at cultigens. The ICNCP does not recognize the use of trade designations and other marketing devices as scientifically acceptable names, but does provide advice on how they should be presented.

Not all cultigens have been given names according to the *Cultivated Plant Code*. Apart from ancient cultigens like those mentioned above there may be occasional anthropogenic plants such as those that are the result of breeding, selection, and tissue grafting that are of no commercial value and have therefore not been given names according to the ICNCP.

Formal definition

A cultigen is a plant whose origin or selection is primarily due to intentional human activity.

The Wild/Cultivated Distinction

Interest in the distinction between wild and cultivated plants dates back to antiquity. Botanical historian Alan Morton notes that wild and cultivated plants (cultigens) were of intense interest to the ancient Greek botanists (partly for religious reasons) and that the distinction was discussed in some detail by Theophrastus (370–285 BCE) the "Father of Botany". Theophrastus was a pupil of both Plato and Aristotle and succeeded the latter as head of the Peripatetic School of Philosophy at the Lyceum in Athens. Theophrastus accepted the view that it was human action, not divine intervention, that produced cultivated plants (cultigens) from wild plants and he also *"had an inkling of the limits of culturally induced (phenotypic) changes and of the importance of genetic constitution"* (*Historia Plantarum* III, 2,2 and *Causa Plantarum* I, 9,3). He also noted that cultivated varieties of fruit trees would degenerate if cultivated from seed.

Origin of Term

Liberty Hyde Bailey 1858-1954, who coined the word *cultigen* in 1918

The word cultigen was coined in 1918 by Liberty Hyde Bailey (1858–1954) an American horticulturist, botanist and cofounder of the American Society for Horticultural Science. He was aware of the need for special categories for those cultivated plants that had arisen by intentional human activity and which would not fit neatly into the Linnaean hierarchical classification of ranks used by the *International Rules of Botanical Nomenclature* (which later became the *International Code of Nomenclature for algae, fungi, and plants*).

In his 1918 paper Bailey noted that for anyone preparing a descriptive account of the cultivated plants of a country (he was at that time preparing such an account for North America) it would be clear that there are two *gentes* or kinds of plants. Firstly, those that are of known origin or nativity "of known habitat". These he referred to as indigens. The other kind was:

" … a domesticated group of which the origin may be unknown or indefinite, which has such characters as to separate it from known indigens, and which is probably not represented by any type specimen or exact description, having therefore no clear taxonomic beginning."

He called this second kind of plant a cultigen, the word derived from the conflation of the Latin *cultus* – cultivated, and *gens* – kind.

In 1923 Bailey extended his original discussion emphasising that he was dealing with plants at the rank of species and he referred to indigens as:

"those that are discovered in the wild"

and cultigens as plants that:

"arise in some way under the hand of man"

He then defined a cultigen as:

"a species, or its equivalent, that has appeared under domestication"

Bailey's Definitions

Bailey soon altered his 1923 definition of cultigen when, in 1924, he gave a new definition in the Glossary of his *Manual of Cultivated Plants* as:

" Plant or group known only in cultivation; presumably originating under domestication; contrast with indigen "

This, in essence, is the definition given at the head of this piece. This definition of the cultigen permits the recognition of cultivars, unlike the 1923 definition which restricts the idea of the cultigen to plants at the rank of species.

In later publications of the *Liberty Hyde Bailey Hortorium*, Cornell, the idea of the cultigen having the rank of species returned (e.g. *Hortus Second* in 1941 and *Hortus Third* in 1976): both of these publications indicate that the terms cultigen and cultivar are not synonymous and that cultigens exist at the rank of species only.

"A cultigen is a plant or group of apparent specific rank, known only in cultivation, with no determined nativity, presumably having originated, in the form in which we know it, under domestication. Compare indigen. Examples are Cucurbita maxima, Phaseolus vulgaris, Zea mays ".

Recent usage in horticulture has, however, maintained a distinction between cultigen and cultivar while nevertheless *allowing the inclusion of cultivars* within the definition.

Cultivars

Cultigen and cultivar may be confused with one another. Cultigen is a general-purpose term encompassing not only plants with cultivar names but others as well, while cultivar is a formal classification category (in the ICNCP).

Although in his 1923 paper Bailey used only the rank of species for the cultigen, it was clear to him that many domesticated plants were more like botanical varieties than species and so he established a new classification category for these, the cultivar, generally assumed to be a contraction of the words "cultivated" and "variety". Bailey was never explicit about the etymology of the word cultivar and it has been suggested that it is a contraction of the words "cultigen" and "variety". He defined cultivar in his 1923 paper as:

... *"a race subordinate to species, that has originated and persisted under cultivation; it is not necessarily, however, referable to a recognised botanical species. It is essentially the equivalent of the botanical variety except in respect to its origin".*

This definition and understanding of cultivar has changed over time.

Usage

Usage in Botany

In botanical literature the word cultigen is generally used to denote a plant which, like the bread wheat (*Triticum aestivum*) is of unknown origin, but presumed to be an ancient human selection. Plants like bread wheat have been given binomials according to the *Botanical Code* and therefore have names with the same form as those of plant species that occur naturally in the wild, but it is not necessary for a cultigen to have a species name, or to have the biological characteristics that distinguish a species. Cultigens can have names at any of various other ranks, including cultivar names, names in the classification categories of grex and group, variety names, forma names, or they may be plants that have been altered by humans (including genetically modified plants) but which have not been given formal names.

Usage in Horticulture

The year 1953 was an important one for cultivated plant taxonomy because this was the date of publication of the first *International Code of Nomenclature for Cultivated Plants* in which Bailey's term cultivar was introduced. It was also the year that the eponymous journal commemorating the work of Bailey (who died in 1954), *Baileya*, was published. In the first volume of *Baileya* taxonomist and colleague of Bailey, George Lawrence, wrote a short article clarifying the distinction between the new term *cultivar* and the *variety*. In the same article he also tried to clarify the critical term *taxon* which had been introduced by German biologist Meyer in the 1920s but had only just been introduced and accepted in botanical circles. This brief article by Lawrence is useful for its insight into the understanding of the meaning of the word cultigen at this time. He opens the article:

In 1918, L.H. Bailey distinguished those plants originating in cultivation from the native plants by designating the former as cultigens and the latter as indigens (indigenous or native to the region). At the same time he proposed the term cultivar to distinguish varieties originating in cultivation from botanical varieties known first in the wild.

In horticulture the definition and use of the terms cultigen and cultivar has varied. One example is the definition given in the Botanical Glossary of *The New Royal Horticultural Society Dictionary of Gardening* which defines cultigen as:

" A plant found only in cultivation or in the wild having escaped from cultivation; included here are many hybrids and cultivars, "...

The use of cultigen in this sense is essentially the same as the definition of the cultigen published by Bailey in 1924.

The *Cultivated Plant Code*, however, states that cultigens are "maintained as recognisable entities solely by continued propagation", and thus would not include plants that have evolved subsequent to escape from cultivation.

Recommended Usage

Wider use of the term cultigen as defined here has been proposed for the following reasons:

- supports current usage in horticulture

- assists clarity in non-technical discussions about "wild" and "cultivated" plants (for example, cultivated plants as commonly understood (plants in cultivation) are not the same as the "cultivated plants" of the *Cultivated Plant Code*, and the distinction between "wild" and "cultivated" habitats is becoming progressively blurred)

- has the potential to simplify the language and definitions used in the Articles and Recommendations of the *Cultivated Plant Code*

- gives greater precision and clarity to the definition of the respective scope, terminology and concepts of the *Botanical Code* and the *Cultivated Plant Code*

- avoids the potential for confusion within the *Cultivated Plant Code* over its scope, that is, whether it is concerned with:

    - where plants are growing (in the wild or in cultivation)

    - how they originated (whether they are the result of intentional human activity or not)

    - whether it simply provides a mechanism for regulating the names of those cultigens requiring special classification categories that are not part of the Linnaean hierarchy of the *Botanical Code* i.e. cultivar and Group names

Critique of Definition

Potential misunderstandings and questions arising from the definition of cultigen given here have been discussed in the literature and are summarised below.

- Natural and *artificial* selection

The selection process is termed "artificial" when human preferences or influences have a significant effect on the evolution of a particular population or species. Note: artificial selection is a part of the overall selection process – it does not imply that humans are not part of nature, it is simply useful sometimes to distinguish when there has been human influence on selection (as with cultigens).

- What exactly does *altered* mean?

There are cases that do not seem to comply with the definition. For example, we can presume that the entire global flora is changing as a result of human-induced climate change. Does this mean that all plants are cultigens?

In cases like this the definition refers to "deliberate" selection and this would be of particular plant characteristics that are not exhibited by a plant's wild counterparts.

- What exactly does *deliberately selected* mean?

From the moment a plant is taken from the wild it is subject to human selection pressure – from the selection of the original propagation material to the purchase of the plant in a nursery. Surely this form of selection is not deliberate? Again, the early human selection of crops 7,000-10,000 years ago is thought to have occurred quite unintentionally. Variants useful to horticulture often arise spontaneously, they are not deliberate products. Are these cases of unintentional, accidental, or unconscious selection?

There certainly appear to be cases where origin or selection of a plant is not "deliberate". However, the long term propagation of plants that have some utility, usually economic or ornamental, can hardly be regarded as unintentional and these plants will, almost without exception, have characteristic(s) that distinguish them from their wild counterparts.

- What about plants selected from the wild?

Plants like *Quercus robur*, Pedunculate or English Oak, *Liquidambar styraciflua*, Sweetgum and *Eucalyptus globulus*, Blue Gum grown in parks and gardens are essentially the same as their wild counterparts and are therefore not cultigens. However, occasionally within natural plant variation there occur characters that are of value to horticulture but of little interest to botany. For example a plant might have flowers of several different colours but these may not have been given formal botanical names. It is customary in horticulture to introduce such variants to commerce and to give them cultivar names. Technically these plants have not been deliberately altered in any way from plants growing (or once growing) in the wild but as they are *deliberately selected* and *named* it seems permissible to refer to them as cultigens. These occurrences are very few. The definition could be (clumsily) extended by mentioning that selection can be for "desirable variation that is not recognised in botanical nomenclature".

- What about gene flow between populations?

Occasionally cultigens escape from cultivation into the wild where they breed with indigenous plants. Selections may be made from the progeny in the wild and brought back into cultivation where they are used for breeding and the results of the breeding again escape into the wild to breed with indigenous plants. *Lantana* has behaved much like this. The genetic material of a cultigen may become part of the gene pool of a population where, over time, it may be largely or completely swamped. In cases like this what plants are to be called cultigens?

Whether a plant is a cultigen or not does not depend on where it is growing. If it complies with the definition then it is a cultigen. Cases like this have always been difficult for botanical nomenclature. Unnamed progeny in the wild might be given a name like *Lantana* aff. *camara* (aff. = having affinities with) or may remain unnamed. Its cultigenic origin may or may not be recognised by the allocation of a cultivar name.

- Plants of unknown origin

Occasionally plants will occur whose origin is unknown. Plants growing in cultivation that are unknown in the wild may be determined as cultigenic as a result of scientific investigation, but may remain a mystery.

- Difficult cases

It may happen that a hybrid cross that has occurred in nature is also performed deliberately in cultivation and that the progeny appear identical. How do we know which plants are cultigens?

If the cross in cultivation is followed by deliberate selection and naming then this will indicate a cultigen. However in a case like this it may not be possible to tell.

Cultivar

Osteospermum 'Pink Whirls'
A cultivar selected for its intriguing and colourful flowers

A cultivar is a plant or grouping of plants selected for desirable characteristics that can be maintained by propagation. Most cultivars have arisen in cultivation but a few are special selections from the wild. Popular ornamental garden plants like roses, camellias, daffodils, rhododendrons, and azaleas are cultivars produced by careful breeding and selection for flower colour and form. Similarly, the world's agricultural food crops are almost exclusively cultivars that have been selected for characteristics such as improved yield, flavour, and resistance to disease: very few wild plants are now used as food sources. Trees used in forestry are also special selections grown for their enhanced quality and yield of timber.

Cultivars form a major part of Liberty Hyde Bailey's broader grouping, the cultigen, defined as a plant whose origin or selection is primarily due to intentional human activity. *Cultivar* was coined by Bailey and it is generally regarded as a portmanteau of "cultivated" and "variety", but could also be derived from "cultigen" and "variety". A cultivar is not the same as a botanical variety, and there are differences in the rules for the formation and use of the names of botanical varieties and cultivars. In recent times the naming of cultivars has been complicated by the use of statutory plant patents and plant breeders' rights names.

The International Union for the Protection of New Varieties of Plants (UPOV – French: *Union internationale pour la protection des obtentions végétales*) offers legal protection of plant cultivars to people or organisations who introduce new cultivars to commerce. UPOV requires that a cultivar be distinct, uniform and stable. To be distinct, it must have characteristics that easily distinguish it from any other known cultivar. To be uniform and stable, the cultivar must retain these characteristics under repeated propagation.

The naming of cultivars is an important aspect of cultivated plant taxonomy, and the correct naming of a cultivar is prescribed by the Rules and Recommendations of the *International Code of Nomenclature for Cultivated Plants* (the *ICNCP*, commonly known as the *Cultivated Plant Code*). A cultivar is given a cultivar name, which consists of the scientific Latin botanical name followed by a cultivar epithet. The cultivar epithet is usually in a vernacular language. For example, the full cultivar name of the King Edward potato is *Solanum tuberosum* 'King Edward'. The 'King Edward' part of the name is the cultivar epithet which, according to the Rules of the *Cultivated Plant Code*, is bounded by single quotation marks.

Origin of Term

The origin of the term "cultivar" arises from the need to distinguish between wild plants and those with characteristics that have arisen in cultivation (what we now call cultigens). This distinction dates back to the Greek philosopher Theophrastus (370–285 BCE), the "Father of Botany", who was keenly aware of this difference. Botanical historian Alan Morton notes that Theophrastus in his *Enquiry into Plants* "*had an inkling of the limits of culturally induced (phenotypic) changes and of the importance of genetic constitution*" (*Historia Plantarum* III, 2,2 and *Causa Plantarum* I, 9,3).

The International Code of Nomenclature for algae, fungi, and plants uses as its starting point for modern botanical nomenclature those Latin names that appeared in Linnaeus' publications *Species Plantarum* (10th ed.) and *Genera Plantarum* (5th ed.). In *Species Plantarum*, Linnaeus (1707–1778) listed all the plants known to him, either directly or from his extensive reading. He recognised the rank of varietas (in English this is the botanical "variety", a rank below that of species and subspecies) and he indicated these varieties by using letters of the Greek alphabet such as α, β, λ in front of the variety name, rather than using the abbreviation var., which is the current convention. Most of the varieties listed by Linnaeus were of "garden" origin rather than being wild plants.

Over time there was an increasing need to distinguish between plants growing in the wild, and those with variations that had been produced in cultivation. In the nineteenth century many "garden-derived" plants were given horticultural names, sometimes in Latin and sometimes in a local language. From about the 1900s, plants produced in cultivation in Europe were recognised in the Scandinavian, Germanic, and Slavic literature through the words *stamm* or *sorte* but these words could not be used internationally since, by international agreement, any new terms had to be based in Latin. In the twentieth century an improved international terminology was proposed for the classification and nomenclature of cultivated plants.

The word *cultivar* was coined in 1923 by Liberty Hyde Bailey of Cornell University, New York State, when he wrote:

The cultigen is a species, or its equivalent, that has appeared under domestication – the plant is cultigenous. I now propose another name, cultivar, for a botanical variety, or for a race subordinate to species, that has originated under cultivation; it is not necessarily, however, referable to a recognized botanical species. It is essentially the equivalent of the botanical variety except in respect to its origin.

In this paper Bailey used only the rank of species for the cultigen but it was clear to him that many domesticated plants were more like botanical varieties than species, and that appears to have motivated the suggestion of the new classification category *cultivar*, which is generally assumed to be a contraction of the words *cultivated* and *variety*. However, Bailey was never explicit about the etymology of the word, and it has been suggested that it is a contraction of the words *cultigen* and *variety*, which seems more appropriate.

The new word *cultivar* was promoted as "euphonious" and "free from ambiguity". It serves a purpose. Its use was subsequently recommended by the first *Cultivated Plant Code*, which was published in 1953, and by 1960 it had achieved wide international acceptance.

Cultigens

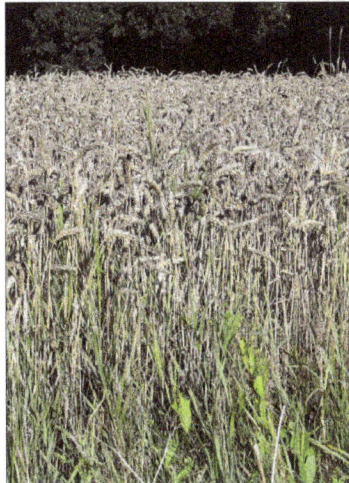

Bread wheat, *Triticum aestivum*, is considered a cultigen, and is a distinct species from other wheats according to the biological species concept. Many separate cultivars have been created within the cultigen. Many other cultigens are not considered to be distinct species, and can be named in other ways.

The terms *cultigen* and *cultivar* may be confused with each other. A *cultigen* is any plant deliberately altered or selected for cultivation by humans; the *Cultivated Plant Code* states that cultigens are "maintained as recognisable entities solely by continued propagation". A cultigen or a component of a cultigen, which is recognisable and stable, can be accepted as a *cultivar*.

Cultigens can have names at any of various ranks, including cultivar names, names in the classification categories of grex and group, binomial Latin species names, variety names, forma names, and they may be plants that have been altered by humans (including genetically modified plants) but which have not been given formal names.

Formal Definition

The *Cultivated Plant Code* notes that the word cultivar is used in two different senses: first, as a "classification category" the cultivar is defined in Article 2 of the *International Code of Nomenclature for Cultivated Plants* (2009, 8th edition) as follows: *The basic category of cultivated plants whose nomenclature is governed by this Code is the cultivar.* There are two other classification categories for cultigens, the grex and the group. The *Code* then defines a *cultivar* as a "taxonomic unit within the classification category of cultivar". This is the sense of *cultivar* that is most generally understood and which is used as a general definition.

A cultivar is an assemblage of plants that (a) has been selected for a particular character or combination of characters, (b) is distinct, uniform and stable in those characters, and (c) when propagated by appropriate means, retains those characters.

Different Kinds

A cultivar of the orchid genus *Oncidium*

Which plants are chosen to be named as cultivars is simply a matter of convenience as the category was created to serve the practical needs of horticulture, agriculture, and forestry.

Members of a particular cultivar are not necessarily genetically identical. The *Cultivated Plant Code* emphasizes that different cultivated plants may be accepted as different cultivars, even if they have the same genome, while cultivated plants with different genomes may be regarded as the same cultivar. The production of cultivars generally entails considerable human involvement although in a few cases it may be as little as simply selecting variation from plants growing in the wild (whether by collecting growing tissue to propagate from or by gathering seed).

Cultivars generally occur as ornamentals and food crops: *Malus* 'Granny Smith' and *Malus* 'Red Delicious' are cultivars of apples propagated by cuttings or grafting, *Lactuca* 'Red Sails' and *Lactuca* 'Great Lakes' are lettuce cultivars propagated by seeds. Named cultivars of *Hosta* and *Hemerocallis* plants are cultivars produced by micropropagation or division.

Clones

Cultivars that are produced asexually are genetically identical and known as clones; this includes plants propagated by division, layering, cuttings, grafts, and budding. The propagating material may be taken from a particular part of the plant, such as a lateral branch, or from a particular phase of the life cycle, such as a juvenile leaf, or from aberrant growth as occurs with witch's broom. Plants whose distinctive characters are derived from the presence of an intracellular organism may also form a cultivar provided the characters are reproduced reliably from generation to generation. Plants of the same chimera (which have mutant tissues close to normal tissue) or graft-chimeras (which have vegetative tissue from different kinds of plants and which originate by grafting) may also constitute a cultivar.

Leucospermum 'Scarlet Ribbon'
A cross performed in Tasmania between *L. glabrum* and *L. tottum*

Seed-produced

Some cultivars "come true from seed", retaining their distinguishing characteristics when grown from seed. Such plants are termed a "variety", "selection" or "strain" but these are ambiguous and confusing words that are best avoided. In general, asexually propagated cultivars grown from seeds produce highly variable seedling plants, and should not be labelled with, or sold under, the parent cultivar's name.

Seed-raised cultivars may be produced by uncontrolled pollination when characteristics that are distinct, uniform and stable are passed from parents to progeny. Some are produced as "lines" that are produced by repeated self-fertilization or inbreeding or "multilines" that are made up of several closely related lines. Sometimes they are F1 hybrids which are the result of a deliberate repeatable single cross between two pure lines. A few F2 hybrid seed cultivars also exist, such as *Achillea* 'Summer Berries'.

Some cultivars are agamospermous plants, which retain their genetic composition and characteristics under reproduction. Occasionally cultivars are raised from seed of a specially selected provenance – for example the seed may be taken from plants that are resistant to a particular disease.

Genetically Modified

Genetically modified plants with characteristics resulting from the deliberate implantation of genetic material from a different germplasm may form a cultivar. However, the International Code of Nomenclature for Cultivated Plants notes, "In practice such an assemblage is often marketed from one or more lines or multilines that have been genetically modified. These lines or multilines often remain in a constant state of development which makes the naming of such an assemblage as a cultivar a futile exercise." However, retired transgenic varieties such as the Fish tomato, which are no longer being developed, do not run into this obstacle and can be given a cultivar name.

Cultivars may be selected because of a change in the ploidy level of a plant which may produce more desirable characteristics.

Cultivar Names

Viola 'Clear Crystals Apricot'
The specific epithet may be omitted from a cultivar name

Every unique cultivar has a unique name within its denomination class (which is almost always the genus). Names of cultivars are regulated by the *International Code of Nomenclature for Cultivated Plants*, and may be registered with an International Cultivar Registration Authority (ICRA). There are sometimes separate registration authorities for different plant types such as roses and camellias. In addition, cultivars may be associated with commercial marketing names referred to in the *Cultivated Plant Code* as "trade designations".

Presenting in Text

A *cultivar name* consists of a botanical name (of a genus, species, infraspecific taxon, interspecific hybrid or intergeneric hybrid) followed by a cultivar epithet. The cultivar epithet is enclosed by single quotes; it should not be italicized if the botanical name is italicized; and each of the words within the epithet is capitalized (with some permitted exceptions such as conjunctions). It is permissible to place a cultivar epithet after a common name provided the common name is botanically unambiguous. Cultivar epithets published before 1 January 1959 were often given a Latin form and can be readily confused with the specific epithets in botanical names; after that date, newly coined cultivar epithets must be in a modern vernacular language to distinguish them from botanical epithets.

- Examples of correct text presentation:

- *Cryptomeria japonica* 'Elegans'

- *Chamaecyparis lawsoniana* 'Aureomarginata' (pre-1959 name, Latin in form)

- *Chamaecyparis lawsoniana* 'Golden Wonder' (post-1959 name, English language)

- *Pinus densiflora* 'Akebono' (post-1959 name, Japanese language)

- Apple 'Sundown'

- Some incorrect text presentation examples:

- *Cryptomeria japonica* "Elegans" (double quotes are unacceptable)

- *Berberis thunbergii* cv. 'Crimson Pygmy' (this once-common usage is now unacceptable, as it is no longer correct to use "cv." in this context; *Berberis thunbergii* 'Crimson Pygmy' is correct)

- *Rosa* cv. 'Peace' (this is now incorrect for two reasons: firstly, the use of "cv."; secondly, "Peace" is a trade designation or "selling name" for the cultivar *R.* 'Madame A. Meilland' and should therefore be printed in a different typeface from the rest of the name, without quote marks, for example: *Rosa* Peace.)

Group Names

Where several very similar cultivars exist they can be associated into a *Group* (formerly *Cultivar-group*). As Group names are used with cultivar names it is necessary to understand their way of presentation. Group names are presented in normal type and the first letter of each word capitalised as for cultivars, but they are not placed in single quotes. When used in a name, the first letter of the word "Group" is itself capitalized.

Presenting in Text

- *Brassica oleracea* Capitata Group (the group of cultivars including all typical cabbages)

- *Brassica oleracea* Botrytis Group (the group of cultivars including all typical cauliflowers)

- *Hydrangea macrophylla* Groupe Hortensia (in French) = *Hydrangea macrophylla* Hortensia Group (in English)

- Where cited with a cultivar name the group should be enclosed in parentheses, as follows:

- *Hydrangea macrophylla* (Hortensia Group) 'Ayesha'

Legal Protection of Cultivars and their Names

Since the 1990s there has been an increasing use of legal protection for newly produced cultivars. Plant breeders expect legal protection for the cultivars they produce. If other growers can immediately propagate and sell these cultivars as soon as they come on the market, the breeder's benefit is largely lost. Legal protection for cultivars is obtained through the use of Plant breeders' rights and plant Patents but the specific legislation and procedures needed to take advantage of this protection vary from country to country.

Controversial use of Legal Protection for Cultivars

The use of legal protection for cultivars can be controversial, particularly for food crops that are staples in developing countries, or for plants selected from the wild and propagated for sale without any additional breeding work; some people consider this practice unethical.

Trade Designations

The formal scientific name of a cultivar, like *Solanum tuberosum* 'King Edward', is a way of uniquely designating a particular kind of plant. This scientific name is in the public domain and cannot be legally protected. Plant retailers wish to maximize their share of the market and one way of doing this is to replace the cumbersome Latin scientific names on plant labels in retail outlets with appealing marketing names that are easy to use, pronounce and remember. Marketing names lie outside the scope of the *Cultivated Plant Code* which refers to them as "trade designations". If a retailer or wholesaler has the sole legal rights to a marketing name then that may offer a sales advantage. Plants protected by Plant breeders' rights (PBR) may have a "true" cultivar name – the recognized scientific name in the public domain, and a "commercial synonym" an additional marketing name that is legally protected: an example would be *Rosa* Fascination = 'Poulmax', the 'Poulmax' being the true scientific name. Because a name that is attractive in one language may have less appeal in another country, a plant may be given different selling names from country to country. Quoting the original cultivar name allows the correct identification of cultivars around the world. The peak body coordinating Plant breeders rights is the Union for the Protection of New Varieties of Plants (*Union internationale pour la protection des obtentions végétales*, UPOV) and this organization maintains a database of new cultivars protected by PBR in all countries.

International Cultivar Registration Authorities

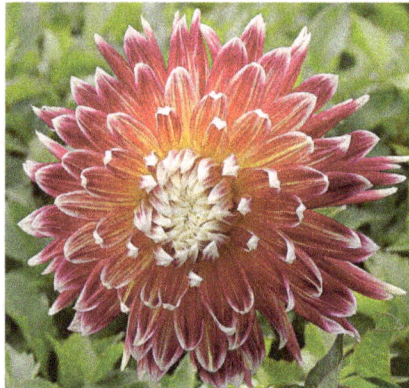

Dahlia 'Akita'
A cultivar selected for flower form and colour

An *International Cultivar Registration Authority* (ICRA) is a voluntary, non-statutory organization appointed by the *Commission for Nomenclature and Cultivar Registration* of the International Society of Horticultural Science. ICRAs are generally formed by societies and institutions specializing in particular plant genera such as *Dahlia* or *Rhododendron* and are currently located in Europe, North America, China, India, Singapore, Australia, New Zealand, South Africa and Puerto Rico.

Each ICRA produces an annual report and its reappointment is considered every four years. The main task is to maintain a register of the names within the group of interest and where possible this is published and placed in the public domain. One major aim is to prevent the duplication of cultivar and Group epithets within a genus, as well as ensuring that names are in accord with the latest edition of the *Cultivated Plant Code*. In this way, over the last 50 years or so, ICRAs have contributed to the stability of cultivated plant nomenclature. In recent times many ICRAs have also recorded trade designations and trademarks used in labelling plant material, to avoid confusion with established names.

New names and other relevant data are collected by and submitted to the ICRA and in most cases there is no cost. The ICRA then checks each new epithet to ensure that it has not been used before and that it conforms with the *Cultivated Plant Code*. Each ICRA also ensures that new names are formally established (i.e. published in hard copy, with a description in a dated publication). They record details about the plant, such as parentage, the names of those concerned with its development and introduction, and a basic description highlighting its distinctive characters. ICRAs are not responsible for assessing the distinctiveness of the plant in question. Most ICRAs can be contacted electronically and many maintain web sites: for an up-to-date listing.

References

- Suzie Key; Julian K-C Ma & Pascal MW Drake (1 June 2008). "Genetically modified plants and human health". Journal of the Royal Society of Medecine. pp. 290–298. Retrieved 11 March 2015.

- Oldeman, l (1994). "The global extent of soil degradation" (PDF). Soil resilience and sustainable land use. 32 (5967): 818–822. Retrieved 7-11-2013.

- Bänziger (2000). "Breeding for drought and nitrogen stress tolerance in maize: from theory to practice". From Theory to Practice: 7–9. Retrieved 7-11-2013.

- Staff (2010). "ISHS :: Commission Nomenclature and Cultivar Registration - International Cultivar Registration Authorities (ICRAs)". ishs.org. Retrieved 5 March 2011.

Important Features of Plant Biology

Chloroplasts are the specialized sub-units in plant cells and mutation is the perpetual modification of the sequence of genome of organisms or a permanent change in their DNAs. The other important features of plant biology are mitochondrion, polyploid and germline. Plant genomics is best understood in confluence with the major topics listed in the following chapter.

Mitochondrion

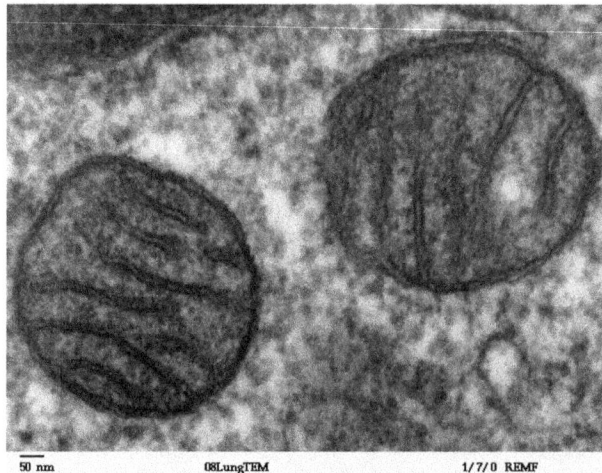

Two mitochondria from mammalian lung tissue displaying their matrix and membranes as shown by electron microscopy

The mitochondrion (plural mitochondria) is a double membrane-bound organelle found in all eukaryotiorganisms, although some cells in some organisms may lack them (e.g. red blood cells). A number of organisms have reduced or transformed their mitochondria into other structures. To date, only one eukaryote, Monocercomonoides, is known to have completely lost its mitochondria. Mitochondria have been described as "the powerhouse of the cell" because they generate most of the cell's supply of adenosine triphosphate (ATP), used as a source of chemical energy.

Mitochondria are commonly between 0.75 and 3μm in diameter but vary considerably in size and structure. Unless specifically stained, they are not visible. In addition to supplying cellular energy, mitochondria are involved in other tasks, such as signaling, cellular differentiation, and cell death, as well as maintaining control of the cell cycle and cell growth. Mitochondrial biogenesis is in turn temporally coordinated with these cellular processes. Mitochondria have been implicated in several human diseases, including mitochondrial disorders, cardiadysfunction, heart failure and autism.

The number of mitochondria in a cell can vary widely by organism, tissue, and cell type. For instance, red blood cells have no mitochondria, whereas liver cells can have more than 2000. The organelle is composed of compartments that carry out specialized functions. These compartments or regions include the outer membrane, the intermembrane space, the inner membrane, and the cristae and matrix. Mitochondrial proteins vary depending on the tissue and the species. In humans, 615 distinct types of protein have been identified from cardiamitochondria, whereas in rats, 940 proteins have been reported. The mitochondrial proteome is thought to be dynamically regulated. Although most of a cell's DNA is contained in the cell nucleus, the mitochondrion has its own independent genome that shows substantial similarity to bacterial genomes.

History

The first observations of intracellular structures that probably represented mitochondria were published in the 1840s. Richard Altmann, in 1890, established them as cell organelles and called them "bioblasts". The term "mitochondria" was coined by Carl Benda in 1898. Leonor Michaelis discovered that Janus green can be used as a supravital stain for mitochondria in 1900. In 1904, Friedrich Meves, made the first recorded observation of mitochondria in plants in cells of the white waterlily, *Nymphaea alba* and in 1908, along with Claudius Regaud, suggested that they contain proteins and lipids. Benjamin F. Kingsbury, in 1912, first related them with cell respiration, but almost exclusively based on morphological observations. In 1913, particles from extracts of guinea-pig liver were linked to respiration by Otto Heinrich Warburg, which he called "grana". Warburg and Heinrich Otto Wieland, who had also postulated a similar particle mechanism, disagreed on the chemical nature of the respiration. It was not until 1925, when David Keilin discovered cytochromes, that the respiratory chain was described.

In 1939, experiments using minced muscle cells demonstrated that cellular respiration using one oxygen atom can form two adenosine triphosphate (ATP) molecules, and, in 1941, the concept of the phosphate bonds of ATP being a form of energy in cellular metabolism was developed by Fritz Albert Lipmann. In the following years, the mechanism behind cellular respiration was further elaborated, although its link to the mitochondria was not known. The introduction of tissue fractionation by Albert Claude allowed mitochondria to be isolated from other cell fractions and biochemical analysis to be conducted on them alone. In 1946, he concluded that cytochrome oxidase and other enzymes responsible for the respiratory chain were isolated to the mitchondria. Eugene Kennedy and Albert Lehninger discovered in 1948 that mitochondria are the site of oxidative phosphorylation in eukaryotes. Over time, the fractionation method was further developed, improving the quality of the mitochondria isolated, and other elements of cell respiration were determined to occur in the mitochondria.

The first high-resolution electron micrographs appeared in 1952, replacing the Janus Green stains as the preferred way of visualising the mitochondria. This led to a more detailed analysis of the structure of the mitochondria, including confirmation that they were surrounded by a membrane. It also showed a second membrane inside the mitochondria that folded up in ridges dividing up the inner chamber and that the size and shape of the mitochondria varied from cell to cell.

The popular term "powerhouse of the cell" was coined by Philip Siekevitz in 1957.

In 1967, it was discovered that mitochondria contained ribosomes. In 1968, methods were developed for mapping the mitochondrial genes, with the genetiand physical map of yeast mitochondrial DNA being completed in 1976.

Origin and Evolution

There are two hypotheses about the origin of mitochondria: endosymbiotiand autogenous. The endosymbiotihypothesis suggests that mitochondria were originally prokaryoticells, capable of implementing oxidative mechanisms that were not possible for eukaryoticells; they became endosymbionts living inside the eukaryote. In the autogenous hypothesis, mitochondria were born by splitting off a portion of DNA from the nucleus of the eukaryoticell at the time of divergence with the prokaryotes; this DNA portion would have been enclosed by membranes, which could not be crossed by proteins. Since mitochondria have many features in common with bacteria, the most accredited theory at present is endosymbiosis.

A mitochondrion contains DNA, which is organized as several copies of a single, circular chromosome. This mitochondrial chromosome contains genes for redox proteins, such as those of the respiratory chain. The CoRR hypothesis proposes that this co-location is required for redox regulation. The mitochondrial genome codes for some RNAs of ribosomes, and the 22 tRNAs necessary for the translation of messenger RNAs into protein. The circular structure is also found in prokaryotes. The proto-mitochondrion was probably closely related to the *Rickettsia*. However, the exact relationship of the ancestor of mitochondria to the alphaproteobacteria and whether the mitochondrion was formed at the same time or after the nucleus, remains controversial.

A recent study by researchers of the University of Hawaii at Manoa and the Oregon State University indicates that the SAR11 clade of bacteria shares a relatively recent common ancestor with the mitochondria existing in most eukaryoticells.

The ribosomes coded for by the mitochondrial DNA are similar to those from bacteria in size and structure. They closely resemble the bacterial 70S ribosome and not the 80S cytoplasmiribosomes, which are coded for by nuclear DNA.

The endosymbiotirelationship of mitochondria with their host cells was popularized by Lynn Margulis. The endosymbiotihypothesis suggests that mitochondria descended from bacteria that somehow survived endocytosis by another cell, and became incorporated into the cytoplasm. The ability of these bacteria to conduct respiration in host cells that had relied on glycolysis and fermentation would have provided a considerable evolutionary advantage. This symbiotirelationship probably developed 1.7 to 2 billion years ago.

A few groups of unicellular eukaryotes have only vestigial mitochondria or derived structures: the microsporidians, metamonads, and archamoebae. These groups appear as the most primitive eukaryotes on phylogenetitrees constructed using rRNA information, which once suggested that they appeared before the origin of mitochondria. However, this is now known to be an artifact of long-branch attraction—they are derived groups and retain genes or organelles derived from mitochondria (e.g., mitosomes and hydrogenosomes).

Monocercomonoides appear to have lost their mitochondria completely and at least some of the mitochondrial functions seem to be carried out by cytoplasmiproteins now.

Structure

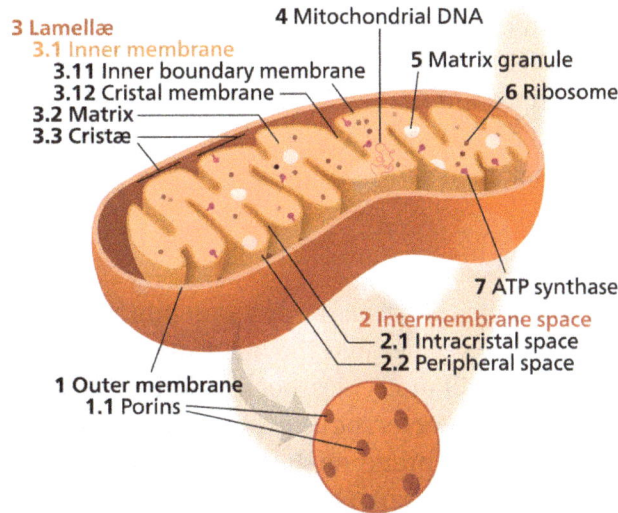

3 Lamellæ
3.1 Inner membrane
  3.11 Inner boundary membrane
  3.12 Cristal membrane
3.2 Matrix
3.3 Cristæ

4 Mitochondrial DNA
5 Matrix granule
6 Ribosome

7 ATP synthase
2 Intermembrane space
  2.1 Intracristal space
  2.2 Peripheral space

1 Outer membrane
  1.1 Porins

Mitochondrion ultrastructure *(interactive diagram)* A mitochondrion has a double membrane; the inner one contains its chemiosmotiapparatus and has deep grooves which increase its surface area. While commonly depicted as an "orange sausage with a blob inside of it" (like it is here), mitochondria can take many shapes and their intermembrane space is quite thin.

A mitochondrion contains outer and inner membranes composed of phospholipid bilayers and proteins. The two membranes have different properties. Because of this double-membraned organization, there are five distinct parts to a mitochondrion. They are:

1. the outer mitochondrial membrane,

2. the intermembrane space (the space between the outer and inner membranes),

3. the inner mitochondrial membrane,

4. the cristae space (formed by infoldings of the inner membrane), and

5. the matrix (space within the inner membrane).

Mitochondria stripped of their outer membrane are called mitoplasts.

Outer Membrane

The outer mitochondrial membrane, which encloses the entire organelle, is 60 to 75 angstroms (Å) thick. It has a protein-to-phospholipid ratio similar to that of the eukaryotiplasma membrane (about 1:1 by weight). It contains large numbers of integral membrane proteins called porins. These porins form channels that allow molecules of 5000 daltons or less in molecular weight to freely diffuse from one side of the membrane to the other. Larger proteins can enter the mitochondrion if a signaling sequence at their N-terminus binds to a large multisubunit protein called translocase of the outer membrane, which then actively moves them across the membrane. Mitochondrial pro-proteins are imported through specialised translocation complexes. The outer membrane also contains enzymes involved in such diverse activities as the elongation of fatty acids, oxidation of epinephrine, and the degradation of tryptophan. These enzymes include monoamine oxidase, rotenone-insensitive NADH-cytochrome c-reductase, kynurenine hydroxylase and fatty acid Co-A ligase. Disruption of the outer membrane permits proteins in the intermembrane space to leak into the cytosol, leading to certain cell death. The mitochondrial outer membrane can associate with the endoplasmireticulum (ER) membrane, in a structure called MAM (mitochondria-associated ER-membrane). This is important in the ER-mitochondria calcium signaling and is involved in the transfer of lipids between the ER and mitochondria. Outside the outer membrane there are small (diameter: 60Å) particles named sub-units of Parson.

Intermembrane Space

The intermembrane space is the space between the outer membrane and the inner membrane. It is also known as perimitochondrial space. Because the outer membrane is freely permeable to small molecules, the concentrations of small molecules, such as ions and sugars, in the intermembrane space is the same as in the cytosol. However, large proteins must have a specifisignaling sequence to be transported across the outer membrane, so the protein composition of this space is different from the protein composition of the cytosol. One protein that is localized to the intermembrane space in this way is cytochrome c.

Inner Membrane

The inner mitochondrial membrane contains proteins with five types of functions:

1. Those that perform the redox reactions of oxidative phosphorylation

2. ATP synthase, which generates ATP in the matrix

3. Specifitransport proteins that regulate metabolite passage into and out of the matrix

4. Protein import machinery

5. Mitochondrial fusion and fission protein

It contains more than 151 different polypeptides, and has a very high protein-to-phospholipid ratio (more than 3:1 by weight, which is about 1 protein for 15 phospholipids). The inner membrane is home to around 1/5 of the total protein in a mitochondrion. In addition, the inner membrane is rich in an unusual phospholipid, cardiolipin. This phospholipid was originally discovered in cow hearts in 1942, and is usually characteristiof mitochondrial and bacterial plasma membranes. Cardiolipin contains four fatty acids rather than two, and may help to make the inner membrane impermeable. Unlike the outer membrane, the inner membrane doesn't contain porins, and is highly impermeable to all molecules. Almost all ions and molecules require special membrane transporters to enter or exit the matrix. Proteins are ferried into the matrix via the translocase of the inner membrane (TIM) complex or via Oxa1. In addition, there is a membrane potential across the inner membrane, formed by the action of the enzymes of the electron transport chain.

Cristae

The inner mitochondrial membrane is compartmentalized into numerous cristae, which expand the surface area of the inner mitochondrial membrane, enhancing its ability to produce ATP. For typical liver mitochondria, the area of the inner membrane is about five times as large as the outer membrane. This ratio is variable and mitochondria from cells that have a greater demand for ATP, such as muscle cells, contain even more cristae. These folds are studded with small round bodies known as F_1 particles or oxysomes. These are not simple random folds but rather invaginations of the inner membrane, which can affect overall chemiosmotifunction.

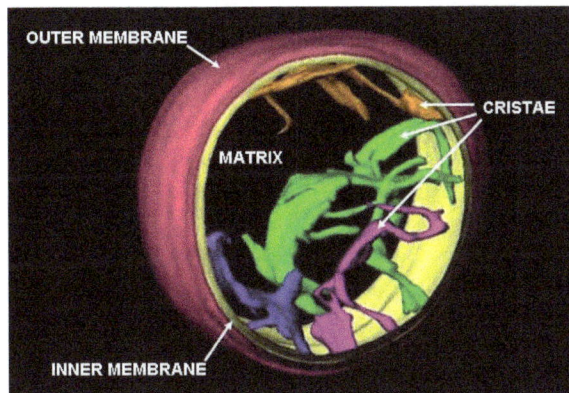

Cross-sectional image of cristae in rat liver mitochondrion to demonstrate the likely 3D structure and relationship to the inner membrane

One recent mathematical modeling study has suggested that the optical properties of the cristae in filamentous mitochondria may affect the generation and propagation of light within the tissue.

Matrix

The matrix is the space enclosed by the inner membrane. It contains about 2/3 of the total protein in a mitochondrion. The matrix is important in the production of ATP with the aid of the ATP synthase contained in the inner membrane. The matrix contains a highly concentrated mixture of hundreds of enzymes, special mitochondrial ribosomes, tRNA, and several copies of the mitochondrial DNA genome. Of the enzymes, the major functions include oxidation of pyruvate and fatty acids, and the citriacid cycle.

Mitochondria have their own genetimaterial, and the machinery to manufacture their own RNAs and proteins. A published human mitochondrial DNA sequence revealed 16,569 base pairs encoding 37 genes: 22 tRNA, 2 rRNA, and 13 peptide genes. The 13 mito-chondrial peptides in humans are integrated into the inner mitochondrial membrane, along with proteins encoded by genes that reside in the host cell's nucleus.

Mitochondria-associated ER Membrane (MAM)

The mitochondria-associated ER membrane (MAM) is another structural element that is increasingly recognized for its critical role in cellular physiology and homeostasis. Once considered a technical snag in cell fractionation techniques, the alleged ER vesicle contaminants that invariably appeared in the mitochondrial fraction have been re-identified as membranous structures derived from the MAM—the interface between mitochondria and the ER. Physical coupling between these two organelles had previously been observed in electron micrographs and has more recently been probed with fluorescence microscopy. Such studies estimate that at the MAM, which may comprise up to 20% of the mitochondrial outer membrane, the ER and mitochondria are separated by a mere 10–25 nm and held together by protein tethering complexes.

Purified MAM from subcellular fractionation has been shown to be enriched in enzymes involved in phospholipid exchange, in addition to channels associated with Ca²⁺ signaling. These hints of a prominent role for the MAM in the regulation of cellular lipid stores and signal transduction have been borne out, with significant implications for mitochondrial-associated cellular phenomena, as discussed below. Not only has the MAM provided insight into the mechanistibasis underlying such physiological processes as intrinsiapoptosis and the propagation of calcium signaling, but it also favors a more refined view of the mitochondria. Though often seen as static, isolated 'powerhouses' hijacked for cellular metabolism through an ancient endosymbiotievent, the evolution of the MAM underscores the extent to which mitochondria have been integrated into overall cellular physiology, with intimate physical and functional coupling to the endomembrane system.

Phospholipid Transfer

The MAM is enriched in enzymes involved in lipid biosynthesis, such as phosphatidylserine synthase on the ER face and phosphatidylserine decarboxylase on the mitochondrial face. Because mitochondria are dynamiorganelles constantly undergoing fission and fusion events, they require a constant and well-regulated supply of phospholipids for membrane integrity. But mitochondria are not only a destination for the phospholipids they finish synthesis of; rather, this organelle also plays a role in inter-organelle trafficking of the intermediates and products of phospholipid biosynthetipathways, ceramide and cholesterol metabolism, and glycosphingolipid anabolism.

Such trafficking capacity depends on the MAM, which has been shown to facilitate transfer of lipid intermediates between organelles. In contrast to the standard vesicular mechanism of lipid transfer, evidence indicates that the physical proximity of the ER and mitochondrial membranes at the MAM allows for lipid flipping between opposed bilayers. Despite this unusual and seemingly energetically unfavorable mechanism, such transport does not require ATP. Instead, in yeast, it has been shown to be dependent on a multiprotein tethering structure termed the ER-mitochondria encounter structure, or ERMES, although it remains unclear whether this structure directly mediates lipid transfer or is required to keep the membranes in sufficiently close proximity to lower the energy barrier for lipid flipping.

The MAM may also be part of the secretory pathway, in addition to its role in intracellular lipid trafficking. In particular, the MAM appears to be an intermediate destination between the rough ER and the Golgi in the pathway that leads to very-low-density lipoprotein, or VLDL, assembly and secretion. The MAM thus serves as a critical metaboliand trafficking hub in lipid metabolism.

Calcium Signaling

A critical role for the ER in calcium signaling was acknowledged before such a role for the mitochondria was widely accepted, in part because the low affinity of Ca²⁺ channels localized to the outer mitochondrial membrane seemed to fly in the face of this organelle's purported responsiveness to changes in intracellular Ca²⁺ flux. But the presence of the MAM resolves this apparent contradiction: the close physical association between the two organelles results in Ca²⁺ microdomains at contact points that facilitate efficient Ca²⁺ transmission from the ER to the mitochondria. Transmission occurs in response to so-called "Ca²⁺ puffs" generated by spontaneous clustering and activation of IP3R, a canonical ER membrane Ca²⁺ channel.

The fate of these puffs—in particular, whether they remain restricted to isolated locales or integrated into Ca^{2+} waves for propagation throughout the cell—is determined in large part by MAM dynamics. Although reuptake of Ca^{2+} by the ER (concomitant with its release) modulates the intensity of the puffs, thus insulating mitochondria to a certain degree from high Ca^{2+} exposure, the MAM often serves as a firewall that essentially buffers Ca^{2+} puffs by acting as a sink into which free ions released into the cytosol can be funneled. This Ca^{2+} tunneling occurs through the low-affinity Ca^{2+} receptor VDAC1, which recently has been shown to be physically tethered to the IP3R clusters on the ER membrane and enriched at the MAM. The ability of mitochondria to serve as a Ca^{2+} sink is a result of the electrochemical gradient generated during oxidative phosphorylation, which makes tunneling of the cation an exergoniprocess. Normally, mild calcium influx from cytosol into the mitochondrial matrix causes transient depolarization that is corrected by pumping out protons.

But transmission of Ca^{2+} is not unidirectional; rather, it is a two-way street. The properties of the Ca^{2+} pump SERCA and the channel IP3R present on the ER membrane facilitate feedback regulation coordinated by MAM function. In particular, the clearance of Ca^{2+} by the MAM allows for spatio-temporal patterning of Ca^{2+} signaling because Ca^{2+} alters IP3R activity in a biphasimanner. SERCA is likewise affected by mitochondrial feedback: uptake of Ca^{2+} by the MAM stimulates ATP production, thus providing energy that enables SERCA to reload the ER with Ca^{2+} for continued Ca^{2+} efflux at the MAM. Thus, the MAM is not a passive buffer for Ca^{2+} puffs; rather it helps modulate further Ca^{2+} signaling through feedback loops that affect ER dynamics.

Regulating ER release of Ca^{2+} at the MAM is especially critical because only a certain window of Ca^{2+} uptake sustains the mitochondria, and consequently the cell, at homeostasis. Sufficient intraorganelle Ca^{2+} signaling is required to stimulate metabolism by activating dehydrogenase enzymes critical to flux through the citriacid cycle. However, once Ca^{2+} signaling in the mitochondria passes a certain threshold, it stimulates the intrinsipathway of apoptosis in part by collapsing the mitochondrial membrane potential required for metabolism. Studies examining the role of pro- and anti-apoptotifactors support this model; for example, the anti-apoptotifactor Bcl-2 has been shown to interact with IP3Rs to reduce Ca^{2+} filling of the ER, leading to reduced efflux at the MAM and preventing collapse of the mitochondrial membrane potential post-apoptotistimuli. Given the need for such fine regulation of Ca^{2+} signaling, it is perhaps unsurprising that dysregulated mitochondrial Ca^{2+} has been implicated in several neurodegenerative diseases, while the catalogue of tumor suppressors includes a few that are enriched at the MAM.

Molecular Basis for Tethering

Recent advances in the identification of the tethers between the mitochondrial and ER membranes suggest that the scaffolding function of the molecular elements involved is secondary to other, non-structural functions. In yeast, ERMES, a multiprotein complex of interacting ER- and mitochondrial-resident membrane proteins, is required for lipid transfer at the MAM and exemplifies this principle. One of its components, for example, is also a constituent of the protein complex required for insertion of transmembrane beta-barrel proteins into the lipid bilayer. However, a homologue of the ERMES complex has not yet been identified in mammalian cells. Other proteins implicated in scaffolding likewise have functions independent of structural tethering at the MAM; for example, ER-resident and mitochondrial-resident mitofusins form heterocomplexes that regulate the number of inter-organelle contact sites, although mitofusins were first identified for their role in fission and fusion events between individual mitochondria. Glucose-related protein 75 (grp75) is another dual-function protein. In addition to the matrix pool of grp75, a portion serves as a chaperone that physically links the mitochondrial and ER Ca^{2+} channels VDAand IP3R for efficient Ca^{2+} transmission at the MAM. Another potential tether is Sigma-1R, a non-opioid receptor whose stabilization of ER-resident IP3R may preserve communication at the MAM during the metabolistress response.

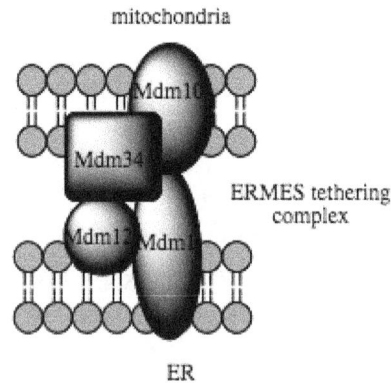

Model of the yeast multimeritethering complex, ERMES

Perspective

The MAM is a critical signaling, metabolic, and trafficking hub in the cell that allows for the integration of ER and mitochondrial physiology. Coupling between these organelles is not simply structural but functional as well and critical for overall cellular physiology and homeostasis. The MAM thus offers a perspective on mitochondria that diverges from the traditional view of this organelle as a static, isolated unit appropriated for its metabolicapacity by the cell. Instead, this mitochondrial-ER interface emphasizes the integration of the mitochondria, the product of an endosymbiotievent, into diverse cellular processes.

Organization and Distribution

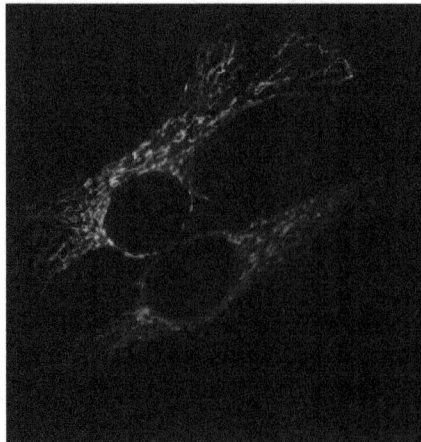

Typical mitochondrial network (green) in two human cells (HeLa cells)

Mitochondria (and related structures) are found in all eukaryotes (except one—the Oxymonad *Monocercomonoides* sp.). Although commonly depicted as bean-like structures they form a highly dynaminetwork in the majority of cells where they constantly undergo fission and fusion. Mitochondria vary in number and location according to cell type. A single mitochondrion is often found in unicellular organisms. Conversely, numerous mitochondria are found in human liver cells, with about 1000–2000 mitochondria per cell, making up 1/5 of the cell volume. The mitochondrial content of otherwise similar cells can vary substantially in size and membrane potential, with differences arising from sources including uneven partitioning at cell divisions, leading to extrinsidifferences in ATP levels and downstream cellular processes. The mitochondria can be found nestled between myofibrils of muscle or wrapped around the sperm flagellum. Often, they form a complex 3D branching network inside the cell with the cytoskeleton. The association with the cytoskeleton determines mitochondrial shape, which can affect the function as well: different structures of the mitochondrial network may afford the population a variety of physical, chemical, and signalling advantages or disadvantages. Mitochondria in cells are always distributed along microtubules and the distribution of these organelles is also correlated with the endoplasmireticulum. Recent evidence suggests that vimentin, one of the components of the cytoskeleton, is also critical to the association with the cytoskeleton.

Function

The most prominent roles of mitochondria are to produce the energy currency of the cell, ATP (i.e., phosphorylation of ADP), through respiration, and to regulate cellular metabolism. The central set of reactions involved in ATP production are collectively known as the citriacid cycle, or the Krebs cycle. However, the mitochondrion has many other functions in addition to the production of ATP.

Energy Conversion

A dominant role for the mitochondria is the production of ATP, as reflected by the large number of proteins in the inner membrane for this task. This is done by oxidizing the major products of glucose: pyruvate, and NADH, which are produced in the cytosol. This type of cellular respiration known as aerobirespiration, is dependent on the presence of oxygen. When oxygen is limited, the glycolytiproducts will be metabolized by anaerobifermentation, a process that is independent of the mitochondria. The production of ATP from glucose has an approximately 13-times higher yield during aerobirespiration compared to fermentation. Recently it has been shown that plant mitochondria can produce a limited amount of ATP without oxygen by using the alternate substrate nitrite. ATP crosses out through the inner membrane with the help of a specifiprotein, and across the outer membrane via porins. ADP returns via the same route.

Pyruvate and the Citriacid Cycle

Pyruvate molecules produced by glycolysis are actively transported across the inner mitochondrial membrane, and into the matrix where they can either be oxidized and combined with coenzyme A to form CO_2, acetyl-CoA, and NADH, or they can be carboxylated (by pyruvate carboxylase) to form oxaloacetate. This latter reaction "fills up" the amount of oxaloacetate in the citriacid cycle, and is therefore an anaplerotireaction, increasing the cycle's capacity to metabolize acetyl-CoA when the tissue's energy needs (e.g. in muscle) are suddenly increased by activity.

In the citriacid cycle, all the intermediates (e.g. citrate, iso-citrate, alpha-ketoglutarate, succinate, fumarate, malate and oxaloacetate) are regenerated during each turn of the cycle. Adding more of any of these intermediates to the mitochondrion therefore means that the additional amount is retained within the cycle, increasing all the other intermediates as one is converted into the other. Hence, the addition of any one of them to the cycle has an anaplerotieffect, and its removal has a cataplerotieffect. These anaplerotiand cataplerotireactions will, during the course of the cycle, increase or decrease the amount of oxaloacetate available to combine with acetyl-CoA to form citriacid. This in turn increases or decreases the rate of ATP production by the mitochondrion, and thus the availability of ATP to the cell.

Acetyl-CoA, on the other hand, derived from pyruvate oxidation, or from the beta-oxidation of fatty acids, is the only fuel to enter the citriacid cycle. With each turn of the cycle one molecule of acetyl-CoA is consumed for every molecule of oxaloacetate present in the mitochondrial matrix, and is never regenerated. It is the oxidation of the acetate portion of acetyl-CoA that produces CO_2 and water, with the energy thus released captured in the form of ATP.

In the liver, the carboxylation of cytosolipyruvate into intra-mitochondrial oxaloacetate is an early step in the gluconeogenipathway, which converts lactate and de-aminated alanine into glucose, under the influence of high levels of glucagon and/or epinephrine in the blood. Here, the addition of oxaloacetate to the mitochondrion does not have a net anaplerotieffect, as another citriacid cycle intermediate (malate) is immediately removed from the mitochondrion to be converted into cyto-solioxaloacetate, which is ultimately converted into glucose, in a process that is almost the reverse of glycolysis.

The enzymes of the citriacid cycle are located in the mitochondrial matrix, with the exception of succinate dehydrogenase, which is bound to the inner mitochondrial membrane as part of Complex II. The citriacid cycle oxidizes the acetyl-CoA to carbon dioxide, and, in the process, produces reduced cofactors (three molecules of NADH and one molecule of $FADH_2$) that are a source of electrons for the *electron transport chain*, and a molecule of GTP (that is readily converted to an ATP).

NADH and $FADH_2$: the Electron Transport Chain

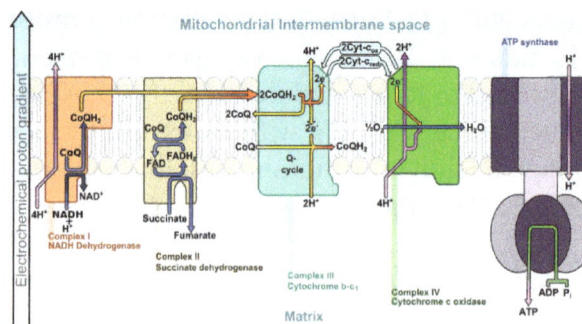

Diagram of the electron transport chain in the mitonchondrial intermembrane space

The redox energy from NADH and $FADH_2$ is transferred to oxygen (O_2) in several steps via the electron transport chain. These energy-rich molecules are produced within the matrix via the citriacid cycle but are also produced in the cytoplasm by glycolysis. Reducing equivalents from the cytoplasm can be imported via the malate-aspartate shuttle system of antiporter proteins or feed into the electron transport chain using a glycerol phosphate shuttle. Protein complexes in the inner membrane (NADH dehydrogenase (ubiquinone), cytochrome reductase, and cytochrome oxidase) perform the transfer and the incremental release of energy is used to pump protons (H⁺) into the intermembrane space. This process is efficient, but a small percentage of electrons may prematurely reduce oxygen, forming reactive oxygen species such as superoxide. This can cause oxidative stress in the mitochondria and may contribute to the decline in mitochondrial function associated with the aging process.

As the proton concentration increases in the intermembrane space, a strong electrochemical gradient is established across the inner membrane. The protons can return to the matrix through the ATP synthase complex, and their potential energy is used to synthesize ATP from ADP and inorganiphosphate (P_i). This process is called chemiosmosis, and was first described by Peter Mitchell who was awarded the 1978 Nobel Prize in Chemistry for his work. Later, part of the 1997 Nobel Prize in Chemistry was awarded to Paul D. Boyer and John E. Walker for their clarification of the working mechanism of ATP synthase.

Heat Production

Under certain conditions, protons can re-enter the mitochondrial matrix without contributing to ATP synthesis. This process is known as *proton leak* or *mitochondrial uncoupling* and is due to the facilitated diffusion of protons into the matrix. The process results in the unharnessed potential energy of the proton electrochemical gradient being released as heat. The process is mediated by a proton channel called thermogenin, or UCP1. Thermogenin is a 33 kDa protein first discovered in 1973. Thermogenin is primarily found in brown adipose tissue, or brown fat, and is responsible for non-shivering thermogenesis. Brown adipose tissue is found in mammals, and is at its highest levels in early life and in hibernating animals. In humans, brown adipose tissue is present at birth and decreases with age.

Storage of Calcium Ions

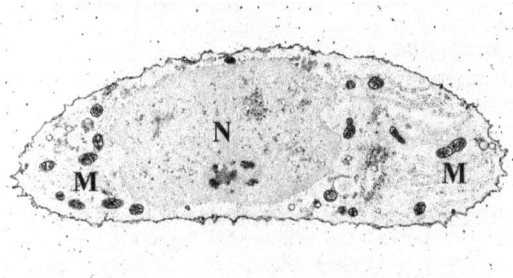

Transmission electron micrograph of a chondrocyte, stained for calcium, showing its nucleus (N) and mitochondria (M).

The concentrations of free calcium in the cell can regulate an array of reactions and is important for signal transduction in the cell. Mitochondria can transiently store calcium, a contributing process for the cell's homeostasis of calcium. In fact, their ability to rapidly take in calcium for later release makes them very good "cytosolibuffers" for calcium. The endoplasmireticulum (ER) is the most significant storage site of calcium, and there is a significant interplay between the mitochondrion and ER with regard to calcium. The calcium is taken up into the matrix by the mitochondrial calcium uniporter on the inner mitochondrial membrane. It is primarily driven by the mitochondrial membrane potential. Release of this calcium back into the cell's interior can occur via a sodium-calcium exchange protein or via "calcium-induced-calcium-release" pathways. This can initiate calcium spikes or calcium waves with large changes in the membrane potential. These can activate a series of second messenger system proteins that can coordinate processes such as neurotransmitter release in nerve cells and release of hormones in endocrine cells.

Ca^{2+} influx to the mitochondrial matrix has recently been implicated as a mechanism to regulate respiratory bioenergetics by allowing the electrochemical potential across the membrane to transiently "pulse" from $\Delta\Psi$-dominated to pH-dominated, facilitating a reduction of oxidative stress. In neurons, concomitant increases in cytosoliand mitochondrial calcium act to synchronize neuronal activity with mitochondrial energy metabolism. Mitochondrial matrix calcium levels can reach the tens of micromolar levels, which is necessary for the activation of isocitrate dehydrogenase, one of the key regulatory enzymes of the Kreb's cycle.

Additional Functions

Mitochondria play a central role in many other metabolitasks, such as:

- Signaling through mitochondrial reactive oxygen species

- Regulation of the membrane potential

- Apoptosis-programmed cell death

- Calcium signaling (including calcium-evoked apoptosis)

- Regulation of cellular metabolism

- Certain heme synthesis reactions

- Steroid synthesis.

- Hormonal signaling Mitochondria are sensitive and responsive to hormones, in part by the action of mitochondrial estrogen receptors (mtERs). These receptors have been found in various tissues and cell types, including brain and heart

Some mitochondrial functions are performed only in specifitypes of cells. For example, mitochondria in liver cells contain enzymes that allow them to detoxify ammonia, a waste product of protein metabolism. A mutation in the genes regulating any of these functions can result in mitochondrial diseases.

Cellular Proliferation Regulation

The relationship between cellular proliferation and mitochondria has been investigated using cervical cancer HeLa cells. Tumor cells require an ample amount of ATP (Adenosine triphosphate) in order to synthesize bioactive compounds such as lipids, proteins, and nucleotides for rapid cell proliferation. The majority of ATP in tumor cells is generated via the oxidative phosphorylation pathway (OxPhos). Interference with OxPhos have shown to cause cell cycle arrest suggesting that mitochondria play a role in cell proliferation. Mitochondrial ATP production is also vital for cell division in addition to other basifunctions in the cell including the regulation of cell volume, solute concentration, and cellular architecture. ATP levels differ at various stages of the cell cycle suggesting that there is a relationship between the abundance of ATP and the cell's ability to enter a new cell cycle. ATP's role in the basifunctions of the cell make the cell cycle sensitive to changes in the availability of mitochondrial derived ATP. The variation in ATP levels at different stages of the cell cycle support the hypothesis that mitochondria play an important role in cell cycle regulation. Although the specifimechanisms between mitochondria and the cell cycle regulation is not well understood, studies have shown that low energy cell cycle checkpoints monitor the energy capability before committing to another round of cell division.

Genome

The 16,569 bp human mitochondrial genome encoding 37 genes, *i.e.*, 28 on the H-strand and 9 on the L-strand.

Mitochondria contain their own genome, an indication that they are derived from bacteria through endoymbiosis. However, the ancestral endosymbiont genome has lost most of its genes so that the mitochondrial genome is one of the most reduced genomes across organisms.

The human mitochondrial genome is a circular DNA molecule of about 16 kilobases. It encodes 37 genes: 13 for subunits of respiratory complexes I, III, IV and V, 22 for mitochondrial tRNA (for the 20 standard amino acids, plus an extra gene for leucine and serine), and 2 for rRNA. One mitochondrion can contain two to ten copies of its DNA.

As in prokaryotes, there is a very high proportion of coding DNA and an absence of repeats. Mitochondrial genes are transcribed as multigenitranscripts, which are cleaved and polyadenylated to yield mature mRNAs. Not all proteins necessary for mitochondrial function are encoded by the mitochondrial genome; most are coded by genes in the cell nucleus and the corresponding proteins are imported into the mitochondrion. The exact number of genes encoded by the nucleus and the mitochondrial genome differs between species. Most mitochondrial genomes are circular, although exceptions have been reported. In general, mitochondrial DNA lacks introns, as is the case in the human mitochondrial genome; however, introns have been observed in some eukaryotimitochondrial DNA, such as that of yeast and protists, including *Dictyostelium discoideum*. Between protein-coding regions, tRNAs are present. During transcription, the tRNAs acquire their characteristiL-shape that gets recognized and cleaved by specifienzymes. Mitochondrial tRNA genes have different sequences from the nuclear tRNAs but lookalikes of mitochondrial tRNAs have been found in the nuclear chromosomes with high sequence similarity.

In animals, the mitochondrial genome is typically a single circular chromosome that is approximately 16 kb long and has 37 genes. The genes, while highly conserved, may vary in location. Curiously, this pattern is not found in the human body louse (*Pediculus humanus*). Instead, this mitochondrial genome is arranged in 18 minicircular chromosomes, each of which is 3–4 kb long and has one to three genes. This pattern is also found in other sucking lice, but not in chewing lice. Recombination has been shown to occur between the minichromosomes. The reason for this difference is not known.

Alternative Geneticode

While slight variations on the standard geneticode had been predicted earlier, none was discovered until 1979, when researchers studying human mitochondrial genes determined that they used an alternative code. Although, the mitochondria of many other eukaryotes, including most plants, use the standard code. Many slight variants have been discovered since, including various alternative mitochondrial codes. Further, the AUA, AUC, and AUU codons are all allowable start codons.

Exceptions to the standard geneticode in mitochondria			
Organism	Codon	Standard	Mitochondria
Mammals	AGA, AGG	Arginine	Stop codon
Invertebrates	AGA, AGG	Arginine	Serine
Fungi	CUA	Leucine	Threonine
All of the above	AUA	Isoleucine	Methionine
	UGA	Stop codon	Tryptophan

Some of these differences should be regarded as pseudo-changes in the geneticode due to the phenomenon of RNA editing, which is common in mitochondria. In higher plants, it was thought that CGG encoded for tryptophan and not arginine; however, the codon in the processed RNA was discovered to be the UGG codon, consistent with the standard geneticode for tryptophan. Of note, the arthropod mitochondrial geneticode has undergone parallel evolution within a phylum, with some organisms uniquely translating AGG to lysine.

Evolution and Diversity

Mitochondrial genomes have far fewer genes than the bacteria from which they are thought to be descended. Although some have been lost altogether, many have been transferred to the nucleus, such as the respiratory complex II protein subunits. This is thought to be relatively common over evolutionary time. A few organisms, such as the *Cryptosporidium*, actually have mitochondria that lack any DNA, presumably because all their genes have been lost or transferred. In *Cryptosporidium*, the mitochondria have an altered ATP generation system that renders the parasite resistant to many classical mitochondrial inhibitors such as cyanide, azide, and atovaquone.

Replication and Inheritance

Mitochondria divide by binary fission, similar to bacterial cell division. The regulation of this division differs between eukaryotes. In many single-celled eukaryotes, their growth and division is linked to the cell cycle. For example, a single mitochondrion may divide synchronously with the nucleus. This division and segregation process must be tightly controlled so that each daughter cell receives at least one mitochondrion. In other eukaryotes (in mammals for example), mitochondria may replicate their DNA and divide mainly in response to the energy needs of the cell, rather than in phase with the cell cycle. When the energy needs of a cell are high, mitochondria grow and divide. When the energy use is low, mitochondria are destroyed or become inactive. In such examples, and in contrast to the situation in many single celled eukaryotes, mitochondria are apparently randomly distributed to the daughter cells during the division of the cytoplasm. Understanding of mitochondrial dynamics, which is described as the balance between mitochondrial fusion and fission, has revealed that functional and structural alterations in mitochondrial morphology are important factors in pathologies associated with several disease conditions.

The hypothesis of mitochondrial binary fission has relied on the visualization by fluorescence microscopy and conventional transmission electron microscopy (TEM). The resolution of fluorescence microscopy(~200 nm) is insufficient to distinguish structural details, such as double mitochondrial membrane in mitochondrial division or even to distinguish individual mitochondria when several are close together. Conventional TEM has also some technical limitations in verifying mitochondrial division. Cryo-electron tomography was recently used to visualize mitochondrial division in frozen hydrated intact cells. It revealed that mitochondria divide by budding.

An individual's mitochondrial genes are not inherited by the same mechanism as nuclear genes. Typically, the mitochondria are inherited from one parent only. In humans, when an egg cell is fertilized by a sperm, the egg nucleus and sperm nucleus each contribute equally to the genetimakeup of the zygote nucleus. In contrast, the mitochondria, and therefore the mitochondrial DNA, usually come from the egg only. The sperm's mitochondria enter the egg, but do not contribute genetiinformation to the embryo. Instead, paternal mitochondria are marked with ubiquitin to select them for later destruction inside the embryo. The egg cell contains relatively few mitochondria, but it is these mitochondria that survive and divide to populate the cells of the adult organism. Mitochondria are, therefore, in most cases inherited only from mothers, a pattern known as maternal inheritance. This mode is seen in most organisms, including the majority of animals. However, mitochondria in some species can sometimes be inherited paternally. This is the norm among certain coniferous plants, although not in pine trees and yews. For Mytilids, paternal inheritance only occurs within males of the species. It has been suggested that it occurs at a very low level in humans. There is a recent suggestion that mitochondria that shorten male lifespan stay in the system because they are inherited only through the mother. By contrast, natural selection weeds out mitochondria that reduce female survival as such mitochondria are less likely to be passed on to the next generation. Therefore, it is suggested that human females and female animals tend to live longer than males. The authors claim that this is a partial explanation.

Uniparental inheritance leads to little opportunity for genetirecombination between different lineages of mitochondria, although a single mitochondrion can contain 2–10 copies of its DNA. For this reason, mitochondrial DNA is usually thought to reproduce by binary fission. What recombination does take place maintains genetiintegrity rather than maintaining diversity. However, there are studies showing evidence of recombination in mitochondrial DNA. It is clear that the enzymes necessary for recombination are present in mammalian cells. Further, evidence suggests that animal mitochondria can undergo recombination. The data are a bit more controversial in humans, although indirect evidence of recombination exists. If recombination does not occur, the whole mitochondrial DNA sequence represents a single haplotype, which makes it useful for studying the evolutionary history of populations.

Entities undergoing uniparental inheritance and with little to no recombination may be expected to be subject to Muller's ratchet, the inexorable accumulation of deleterious mutations until functionality is lost. Animal populations of mitochondria avoid this buildup through a developmental process known as the mtDNA bottleneck. The bottleneck exploits stochastiprocesses in the cell to increase in the cell-to-cell variability in mutant load as an organism develops: a single egg cell with some proportion of mutant mtDNA thus produces an embryo where different cells have different mutant loads. Cell-level selection may then act to remove those cells with more mutant mtDNA, leading to a stabilisation or reduction in mutant load between generations. The mechanism underlying the bottleneck is debated, with a recent mathematical and experimental metastudy providing evidence for a combination of random partitioning of mtDNAs at cell divisions and random turnover of mtDNA molecules within the cell.

Population Genetistudies

The near-absence of genetirecombination in mitochondrial DNA makes it a useful source of information for scientists involved in population genetics and evolutionary biology. Because all the mitochondrial DNA is inherited as a single unit, or haplotype, the relationships between mitochondrial DNA from different individuals can be represented as a gene tree. Patterns in these gene trees can be used to infer the evolutionary history of populations. The classiexample of this is in human evolutionary genetics, where the molecular clock can be used to provide a recent date for mitochondrial Eve. This is often interpreted as strong support for a recent modern human expansion out of Africa. Another human example is the sequencing of mitochondrial DNA from Neanderthal bones. The relatively large evolutionary distance between the mitochondrial DNA sequences of Neanderthals and living humans has been interpreted as evidence for the lack of interbreeding between Neanderthals and anatomically modern humans.

However, mitochondrial DNA reflects only the history of the females in a population and so may not represent the history of the population as a whole. This can be partially overcome by the use of paternal genetisequences, such as the non-recombining region of the Y-chromosome. In a broader sense, only studies that also include nuclear DNA can provide a comprehensive evolutionary history of a population.

Recent measurements of the molecular clock for mitochondrial DNA reported a value of 1 mutation every 7884 years dating back to the most recent common ancestor of humans and apes, which is consistent with estimates of mutation rates of autosomal DNA (10^{-8} per base per generation).

Dysfunction and Disease

Mitochondrial Diseases

Damage and subsequent dysfunction in mitochondria is an important factor in a range of human diseases due to their influence in cell metabolism. Mitochondrial disorders often present themselves as neurological disorders, including autism. They can also manifest as myopathy, diabetes, multiple endocrinopathy, and a variety of other systemidisorders. Diseases caused by mutation in the mtDNA include Kearns-Sayre syndrome, MELAS syndrome and Leber's hereditary optineuropathy. In the vast majority of cases, these diseases are transmitted by a female to her children, as the zygote derives its mitochondria and hence its mtDNA from the ovum. Diseases such as Kearns-Sayre syndrome, Pearson syndrome, and progressive external ophthalmoplegia are thought to be due to large-scale mtDNA rearrangements, whereas other diseases such as MELAS syndrome, Leber's hereditary optineuropathy, myocloniepilepsy with ragged red fibers (MERRF), and others are due to point mutations in mtDNA.

In other diseases, defects in nuclear genes lead to dysfunction of mitochondrial proteins. This is the case in Friedreich's ataxia, hereditary spastiparaplegia, and Wilson's disease. These diseases are inherited in a dominance relationship, as applies to most other genetidiseases. A variety of disorders can be caused by nuclear mutations of oxidative phosphorylation enzymes, such as coenzyme Q10 deficiency and Barth syndrome. Environmental influences may interact with hereditary predispositions and cause mitochondrial disease. For example, there may be a link between pesticide exposure and the later onset of Parkinson's disease. Other pathologies with etiology involving mitochondrial dysfunction include schizophrenia, bipolar disorder, dementia, Alzheimer's disease, Parkinson's disease, epilepsy, stroke, cardiovascular disease, chronifatigue syndrome, retinitis pigmentosa, and diabetes mellitus.

Mitochondria-mediated oxidative stress plays a role in cardiomyopathy in Type 2 diabetics. Increased fatty acid delivery to the heart increases fatty acid uptake by cardiomyocytes, resulting in increased fatty acid oxidation in these cells. This process increases the reducing equivalents available to the electron transport chain of the mitochondria, ultimately increasing reactive oxygen species (ROS) production. ROS increases uncoupling proteins (UCPs) and potentiate proton leakage through the adenine nucleotide translocator (ANT), the combination of which uncouples the mitochondria. Uncoupling then increases oxygen consumption by the mitochondria, compounding the increase in fatty acid oxidation. This creates a vicious cycle of uncoupling; furthermore, even though oxygen consumption increases, ATP synthesis does not increase proportionally because the mitochondria is uncoupled. Less ATP availability ultimately results in an energy deficit presenting as reduced cardiaefficiency and contractile dysfunction. To compound the problem, impaired sarcoplasmireticulum calcium release and reduced mitochondrial reuptake limits peak cytosolilevels of the important signaling ion during muscle contraction. The decreased intra-mitochondrial calcium concentration increases dehydrogenase activation and ATP synthesis. So in addition to lower ATP synthesis due to fatty acid oxidation, ATP synthesis is impaired by poor calcium signaling as well, causing cardiaproblems for diabetics.

Possible Relationships to Aging

Given the role of mitochondria as the cell's powerhouse, there may be some leakage of the high-energy electrons in the respiratory chain to form reactive oxygen species. This was thought to result in significant oxidative stress in the mitochondria with high mutation rates of mitochondrial DNA (mtDNA). Hypothesized links between aging and oxidative stress are not new and were proposed in 1956, which was later refined into the mitochondrial free radical theory of aging. A vicious cycle was thought to occur, as oxidative stress leads to mitochondrial DNA mutations, which can lead to enzymatiabnormalities and further oxidative stress.

A number of changes can occur to mitochondria during the aging process. Tissues from elderly patients show a decrease in enzymatiactivity of the proteins of the respiratory chain. However, mutated mtDNA can only be found in about 0.2% of very old cells. Large deletions in the mitochondrial genome have been hypothesized to lead to high levels of oxidative stress and neuronal death in Parkinson's disease.

In Popular Culture

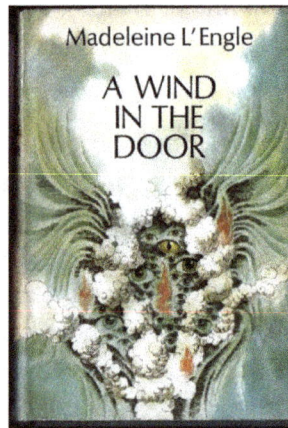

The science fiction novels *A Wind in the Door* and *Parasite Eve* include fictional depictions of mitochondria as major plot elements.

Madeleine L'Engle's 1973 science fantasy novel *A Wind in the Door* prominently features the mitochondria of main character Charles Wallace Murry, as being inhabited by creatures known as the farandolae. The novel also features other characters travelling inside one of Murry's mitochondria.

The 1995 horror fiction novel *Parasite Eve* by Hideaki Sena depicts mitochondria as having some consciousness and mind control abilities, attempting to use these to overtake eukaryotes as the dominant life form. This text was adapted into an eponymous film, video game, and video game sequel all involving a similar premise.

In the *Star Wars* franchise, microorganisms referred to as "midi-chlorians" give some characters the ability to sense and use the Force. George Lucas, director of the 1999 film *Star Wars: Episode I – The Phantom Menace*, in which midi-chlorians were introduced, described them as "a loose depiction of mitochondria". The non-fictional *Midichloria* genus of bacteria was later named after the midi-chlorians of *Star Wars*.

As a result of the mitochondrion's prominence in modern science education, the phrase "mitochondria is [sic] the powerhouse of the cell" became a popular Internet meme. The meme is used to imply that secondary education places an insufficient focus on life skills, compared to academiknowledge such as the role of the mitochondrion, which has been considered comparatively impractical.

Chloroplast

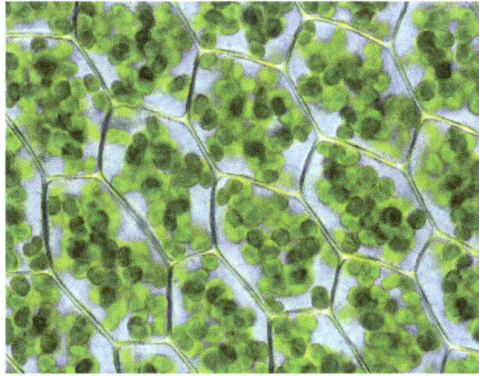

Chloroplasts visible in the cells of *Plagiomnium affine*, the many-fruited thyme moss

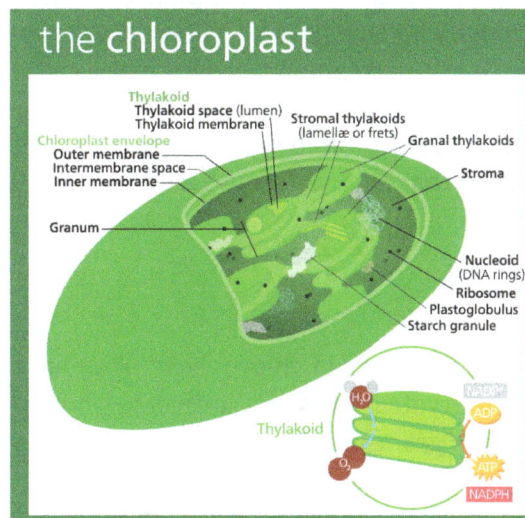

Structure of a typical higher-plant chloroplast

Chloroplasts are organelles, specialized subunits, in plant and algal cells. Their discovery inside plant cells is usually credited to Julius von Sachs (1832–1897), an influential botanist and author of standard botanical textbooks – sometimes called "The Father of Plant Physiology".

Chloroplasts' main role is to conduct photosynthesis, where the photosynthetipigment chlorophyll captures the energy from sunlight and converts it and stores it in the energy-storage molecules ATP and NADPH while freeing oxygen from water. They then use the ATP and NADPH to make organimolecules from carbon dioxide in a process known as the Calvin cycle. Chloroplasts carry out a number of other functions, including fatty acid synthesis, much amino acid synthesis, and the immune response in plants. The number of chloroplasts per cell varies from 1 in algae up to 100 in plants like Arabidopsis and wheat.

A chloroplast is one of three types of plastids, characterized by its high concentration of chlorophyll, the other two types, the leucoplast and the chromoplast, contain little chlorophyll and do not carry out photosynthesis.

Chloroplasts are highly dynamic—they circulate and are moved around within plant cells, and occasionally pinch in two to reproduce. Their behavior is strongly influenced by environmental factors like light color and intensity. Chloroplasts, like mitochondria, contain their own DNA, which is thought to be inherited from their ancestor—a photosyntheticyanobacterium that was engulfed by an early eukaryoticell. Chloroplasts cannot be made by the plant cell and must be inherited by each daughter cell during cell division.

With one exception (the amoeboid *Paulinella chromatophora*), all chloroplasts can probably be traced back to a single endosymbiotievent, when a cyanobacterium was engulfed by the eukaryote. Despite this, chloroplasts can be found in an extremely wide set of organisms, some not even directly related to each other—a consequence of many secondary and even tertiary endosymbiotievents.

Chloroplast Lineages and Evolution

Chloroplasts are one of many types of organelles in the plant cell. They are considered to have originated from cyanobacteria through endosymbiosis—when a eukaryoticell engulfed a photosynthesizing cyanobacterium that became a permanent resident in the cell. Mitochondria are thought to have come from a similar event, where an aerobiprokaryote was engulfed. This origin of chloroplasts was first suggested by the Russian biologist Konstantin Mereschkowski in 1905 after Andreas Schimper observed in 1883 that chloroplasts closely resemble cyanobacteria. Chloroplasts are only found in plants and algae.

Cyanobacterial Ancestor

Cyanobacteria are considered the ancestors of chloroplasts. They are sometimes called blue-green algae even though they are prokaryotes. They are a diverse phylum of bacteria capable of carrying out photosynthesis, and are gram-negative, meaning that they have two cell membranes. Cyanobacteria also contain a peptidoglycan cell wall, which is thicker than in other gram-negative bacteria, and which is located between their two cell membranes. Like chloroplasts, they have thylakoids within. On the thylakoid membranes are photosynthetipigments, including chlorophyll *a*. Phycobilins are also common cyanobacterial pigments, usually organized into hemispherical phycobilisomes attached to the outside of the thylakoid membranes (phycobilins are not shared with all chloroplasts though).

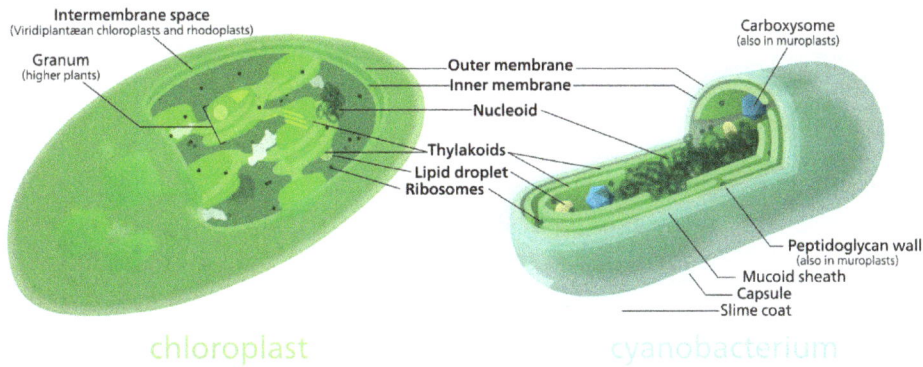

Both chloroplasts and cyanobacteria have a double membrane, DNA, ribosomes, and thylakoids. Both the chloroplast and cyanobacterium depicted are idealized versions (the chloroplast is that of a higher plant)—a lot of diversity exists among chloroplasts and cyanobacteria.

Primary Endosymbiosis

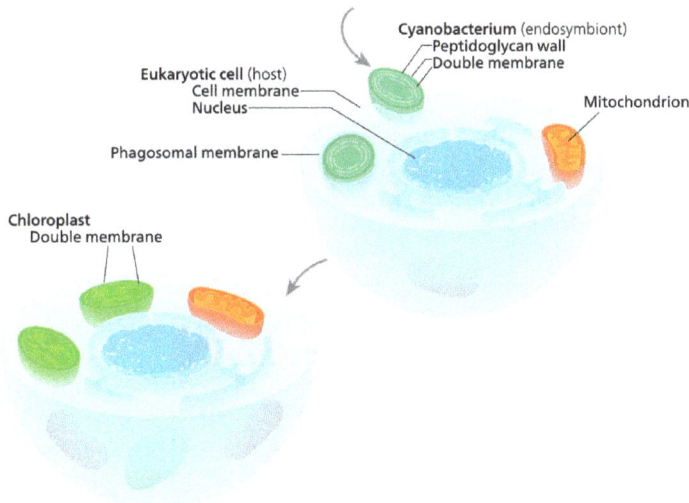

Primary endosymbiosis

A eukaryote with mitochondria engulfed a cyanobacterium in an event of serial primary endosymbiosis, creating a lineage of cells with both organelles. It is important to note that the cyanobacterial endosymbiont already had a double membrane—the phagosomal vacuole-derived membrane was lost.

Somewhere around a billion years ago, a free-living cyanobacterium entered an early eukaryoticell, either as food or as an internal parasite, but managed to escape the phagocytivacuole it was contained in. The two innermost lipid-bilayer membranes that surround all chloroplasts correspond to the outer and inner membranes of the ancestral cyanobacterium's gram negative cell wall, and not the phagosomal membrane from the host, which was probably lost. The new cellular resident quickly became an advantage, providing food for the eukaryotihost, which allowed it to live within it. Over time, the cyanobacterium was assimilated, and many of its genes were lost or transferred to the nucleus of the host. Some of its proteins were then synthesized in the cytoplasm of the host cell, and imported back into the chloroplast (formerly the cyanobacterium).

This event is called *endosymbiosis*, or "cell living inside another cell". The cell living inside the other cell is called the *endosymbiont*; the endosymbiont is found inside the *host cell*.

Chloroplasts are believed to have arisen after mitochondria, since all eukaryotes contain mitochondria, but not all have chloroplasts. This is called *serial endosymbiosis*—an early eukaryote engulfing the mitochondrion ancestor, and some descendants of it then engulfing the chloroplast ancestor, creating a cell with both chloroplasts and mitochondria.

Whether or not chloroplasts came from a single endosymbiotievent, or many independent engulfments across various eukaryotilineages, has been long debated, but it is now generally held that all organisms with chloroplasts either share a single ancestor or obtained their chloroplast from organisms that share a common ancestor that took in a cyanobacterium 600–1600 million years ago.

These chloroplasts, which can be traced back directly to a cyanobacterial ancestor, are known as *primary plastids* (*"plastid"* in this context means almost the same thing as chloroplast). All primary chloroplasts belong to one of three chloroplast lineages—the glaucophyte chloroplast lineage, the rhodophyte, or red algal chloroplast lineage, or the chloroplastidan, or green chloroplast lineage. The second two are the largest, and the green chloroplast lineage is the one that contains the land plants.

Primary endosymbiosis	Secondary endosymbiosis	Tertiary endosymbiosis
Glaucophyta		Chloroplast lineages. A primary endosymbiosis event gave rise to three main lineages of chloroplasts in the glaucophytes, chlorophyta, and rhodophyta. Some of these algae were subsequently engulfed by other algae, becoming secondary (or tertiary) endosymbionts.
Chloroplastida	Euglenophyta	
Land plants	Chlorarachniophyta	
Green algae	Green algal dinophytes	
	Apicomplexa [a]	[a] The apicomplexans (malaria parasites), contain a red algal endosymbiont with a non-photosynthetichloroplast. [b] 2–3 chloroplast membranes [a] 2–4 chloroplast membranes
	Peridinin-type dinophytes [b]	
Rhodophyceae (Red algae)	Cryptophyta	
	Haptophyta	Haptophyte dinophytes [c]
	Heterokontophyta	Diatom dinophytes

Glaucophyta

The alga *Cyanophora*, a glaucophyte, is thought to be one of the first organisms to contain a chloroplast. The glaucophyte chloroplast group is the smallest of the three primary chloroplast lineages, being found in only 13 species, and is thought to be the one that branched off the earliest. Glaucophytes have chloroplasts that retain a peptidoglycan wall between their double membranes, like their cyanobacterial parent. For this reason, glaucophyte chloroplasts are also known as *muroplasts*. Glaucophyte chloroplasts also contain concentriunstacked thylakoids, which surround a carboxysome - an icosahedral structure that glaucophyte chloroplasts and cyanobacteria keep their carbon fixation enzyme rubisco in. The starch that they synthesize collects outside the chloroplast.

Like cyanobacteria, glaucophyte chloroplast thylakoids are studded with light collecting structures called phycobilisomes. For these reasons, glaucophyte chloroplasts are considered a primitive intermediate between cyanobacteria and the more evolved chloroplasts in red algae and plants.

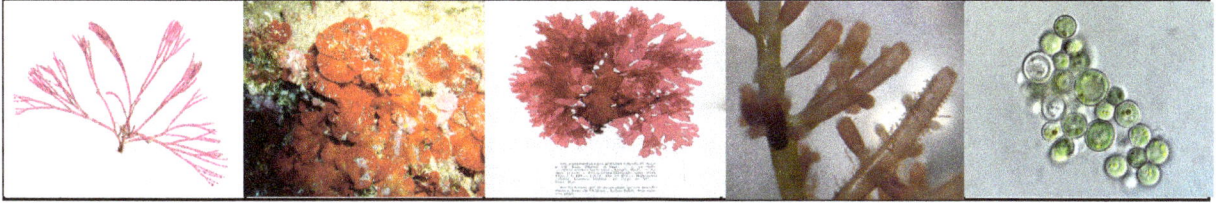

Diversity of red algae Clockwise from top left: *Bornetia secundiflora*, *Peyssonnelia squamaria*, *Cyanidium*, *Laurencia*, *Callophyllis laciniata*. Red algal chloroplasts are characterized by phycobilin pigments which often give them their reddish color.

Rhodophyceae (Red Algae)

The rhodophyte, or red algal chloroplast group is another large and diverse chloroplast lineage. Rhodophyte chloroplasts are also called *rhodoplasts*, literally "red chloroplasts".

Rhodoplasts have a double membrane with an intermembrane space and phycobilin pigments organized into phycobilisomes on the thylakoid membranes, preventing their thylakoids from stacking. Some contain pyrenoids. Rhodoplasts have chlorophyll *a* and phycobilins for photosynthetipigments; the phycobilin phycoerytherin is responsible for giving many red algae their distinctive red color. However, since they also contain the blue-green chlorophyll *a* and other pigments, many are reddish to purple from the combination. The red phycoerytherin pigment is an adaptation to help red algae catch more sunlight in deep water—as such, some red algae that live in shallow water have less phycoerytherin in their rhodoplasts, and can appear more greenish. Rhodoplasts synthesize a form of starch called floridean starch, which collects into granules outside the rhodoplast, in the cytoplasm of the red alga.

Chloroplastida (Green Algae and Plants)

Diversity of green algae Clockwise from top left: *Scenedesmus*, *Micrasterias*, *Hydrodictyon*, *Stigeoclonium*, *Volvox*. Green algal chloroplasts are characterized by their pigments chlorophyll *a* and chlorophyll *b* which give them their green color.

The chloroplastidan chloroplasts, or green chloroplasts, are another large, highly diverse primary chloroplast lineage. Their host organisms are commonly known as the green algae and land plants. They differ from glaucophyte and red algal chloroplasts in that they have lost their phycobilisomes, and contain chlorophyll *b* instead. Most green chloroplasts are (obviously) green, though some aren't, like some forms of *Hæmatococcus pluvialis*, due to accessory pigments that override the chlorophylls' green colors. Chloroplastidan chloroplasts have lost the peptidoglycan wall between their double membrane, and have replaced it with an intermembrane space. Some plants seem to have kept the genes for the synthesis of the peptidoglycan layer, though they've been repurposed for use in chloroplast division instead.

Green algae and plants keep their starch *inside* their chloroplasts, and in plants and some algae, the chloroplast thylakoids are arranged in grana stacks. Some green algal chloroplasts contain a structure called a pyrenoid, which is functionally similar to the glaucophyte carboxysome in that it is where rubisco and CO_2 are concentrated in the chloroplast.

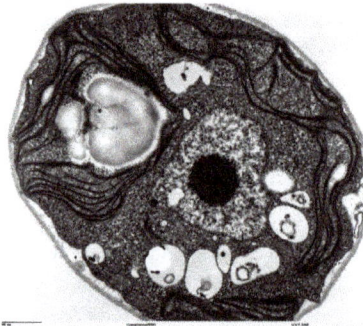

Transmission electron micrograph of *Chlamydomonas reinhardtii*, a green alga that contains a pyrenoid surrounded by starch.

Helicosporidium

Helicosporidium is a genus of nonphotosynthetiparasitigreen algae that is thought to contain a vestigial chloroplast. Genes from a chloroplast and nuclear genes indicating the presence of a chloroplast have been found in Helicosporidium even if nobody's seen the chloroplast itself.

Secondary and Tertiary Endosymbiosis

Many other organisms obtained chloroplasts from the primary chloroplast lineages through secondary endosymbiosis—engulfing a red or green alga that contained a chloroplast. These chloroplasts are known as secondary plastids.

While primary chloroplasts have a double membrane from their cyanobacterial ancestor, secondary chloroplasts have additional membranes outside of the original two, as a result of the secondary endosymbiotievent, when a nonphotosynthetieukaryote engulfed a chloroplast-containing alga but failed to digest it—much like the cyanobacterium at the beginning of this story. The engulfed alga was broken down, leaving only its chloroplast, and sometimes its cell membrane and nucleus, forming a chloroplast with three or four membranes—the two cyanobacterial membranes, sometimes the eaten alga's cell membrane, and the phagosomal vacuole from the host's cell membrane.

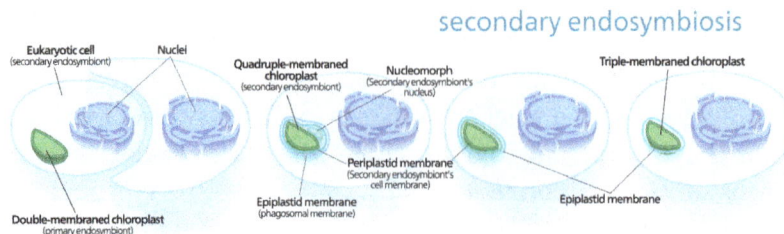

Secondary endosymbiosis consisted of a eukaryotialga being engulfed by another eukaryote, forming a chloroplast with three or four membranes.

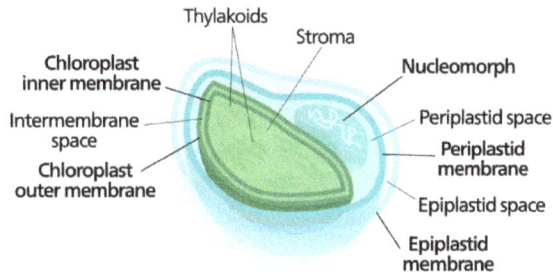

Diagram of a four membraned chloroplast containing a nucleomorph.

The genes in the phagocytosed eukaryote's nucleus are often transferred to the secondary host's nucleus. Cryptomonads and chlorarachniophytes retain the phagocytosed eukaryote's nucleus, an object called a nucleomorph, located between the second and third membranes of the chloroplast.

All secondary chloroplasts come from green and red algae—no secondary chloroplasts from glaucophytes have been observed, probably because glaucophytes are relatively rare in nature, making them less likely to have been taken up by another eukaryote.

Green Algal Derived Chloroplasts

Green algae have been taken up by the euglenids, chlorarachniophytes, a lineage of dinoflagellates, and possibly the ancestor of the chromalveolates in three or four separate engulfments. Many green algal derived chloroplasts contain pyrenoids, but unlike chloroplasts in their green algal ancestors, starch collects in granules outside the chloroplast.

Euglena, a euglenophyte, contains secondary chloroplasts from green algae.

Euglenophytes

Euglenophytes are a group of common flagellated protists that contain chloroplasts derived from a green alga. Euglenophyte chloroplasts have three membranes—it is thought that the membrane of the primary endosymbiont was lost, leaving the cyanobacterial membranes, and the secondary host's phagosomal membrane. Euglenophyte chloroplasts have a pyrenoid and thylakoids stacked in groups of three. Starch is stored in the form of paramylon, which is contained in membrane-bound granules in the cytoplasm of the euglenophyte.

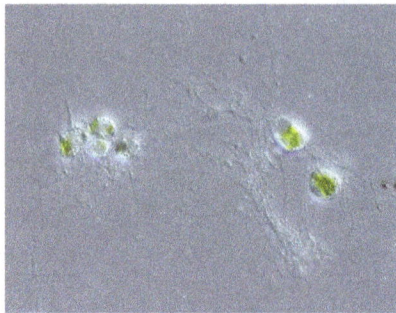

Chlorarachnion reptans is a chlorarachniophyte. Chlorarachniophytes replaced their original red algal endosymbiont with a green alga.

Chlorarachniophytes

Chlorarachniophytes are a rare group of organisms that also contain chloroplasts derived from green algae, though their story is more complicated than that of the euglenophytes. The ancestor of chlorarachniophytes is thought to have been a chromalveolate, a eukaryote with a *red* algal derived chloroplast. It is then thought to have lost its first red algal chloroplast, and later engulfed a green alga, giving it its second, green algal derived chloroplast.

Chlorarachniophyte chloroplasts are bounded by four membranes, except near the cell membrane, where the chloroplast membranes fuse into a double membrane. Their thylakoids are arranged in loose stacks of three. Chlorarachniophytes have a form of starch called chrysolaminarin, which they store in the cytoplasm, often collected around the chloroplast pyrenoid, which bulges into the cytoplasm.

Chlorarachniophyte chloroplasts are notable because the green alga they are derived from has not been completely broken down—its nucleus still persists as a nucleomorph found between the second and third chloroplast membranes—the periplastid space, which corresponds to the green alga's cytoplasm.

Early Chromalveolates

Recent research has suggested that the ancestor of the chromalveolates acquired a green algal prasinophyte endosymbiont. The green algal derived chloroplast was lost and replaced with a red algal derived chloroplast, but not before contributing some of its genes to the early chromalveolate's nucleus. The presence of both green algal and red algal genes in chromalveolates probably helps them thrive under fluctuating light conditions.

Red Algal Derived Chloroplasts (Chromalveolate Chloroplasts)

Like green algae, red algae have also been taken up in secondary endosymbiosis, though it is thought that all red algal derived chloroplasts are descended from a single red alga that was engulfed by an early chromalveolate, giving rise to the chromalveolates, some of which, like the ciliates, subsequently lost the chloroplast. This is still debated though.

Pyrenoids and stacked thylakoids are common in chromalveolate chloroplasts, and the outermost membrane of many are continuous with the rough endoplasmireticulum and studded with ribosomes. They have lost their phycobilisomes and exchanged them for chlorophyll *c*, which isn't found in primary red algal chloroplasts themselves.

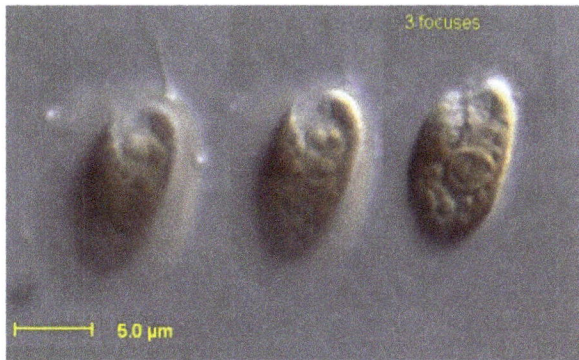

Rhodomonas salina is a cryptophyte.

Cryptophytes

Cryptophytes, or cryptomonads are a group of algae that contain a red-algal derived chloroplast. Cryptophyte chloroplasts contain a nucleomorph that superficially resembles that of the chlorarachniophytes. Cryptophyte chloroplasts have four membranes, the outermost of which is continuous with the rough endoplasmireticulum. They synthesize ordinary starch, which is stored in granules found in the periplastid space—outside the original double membrane, in the place that corresponds to the red alga's cytoplasm. Inside cryptophyte chloroplasts is a pyrenoid and thylakoids in stacks of two.

Their chloroplasts do not have phycobilisomes, but they do have phycobilin pigments which they keep in their thylakoid space, rather than anchored on the outside of their thylakoid membranes.

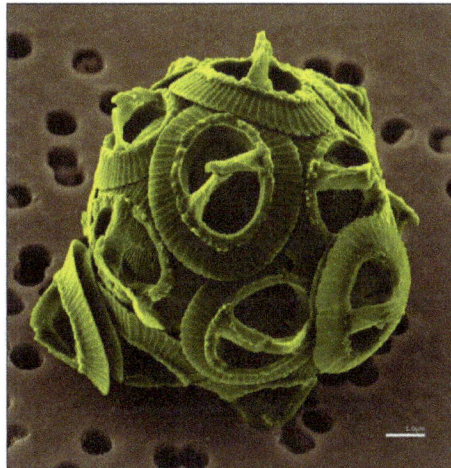

Scanning electron micrograph of *Gephyrocapsa oceanica*, a haptophyte.

Haptophytes

Haptophytes are similar and closely related to cryptophytes, and are thought to be the first chromalveolates to branch off. Their chloroplasts lack a nucleomorph, their thylakoids are in stacks of three, and they synthesize chrysolaminarin sugar, which they store completely outside of the chloroplast, in the cytoplasm of the haptophyte.

Heterokontophytes (Stramenopiles)

The photosynthetipigments present in their chloroplasts give diatoms a greenish-brown color.

The heterokontophytes, also known as the stramenopiles, are a very large and diverse group of algae that also contain red algal derived chloroplasts. Heterokonts include the diatoms and the brown algae, golden algae, and yellow-green algae.

Heterokont chloroplasts are very similar to haptophyte chloroplasts, containing a pyrenoid, triplet thylakoids, and with some exceptions, having an epiplastid membrane connected to the endoplasmireticulum. Like haptophytes, heterokontophytes store sugar in chrysolaminarin granules in the cytoplasm. Heterokontophyte chloroplasts contain chlorophyll a and with a few exceptions chlorophyll c, but also have carotenoids which give them their many colors.

Apicomplexans

Apicomplexans are another group of chromalveolates. Like the helicosproidia, they're parasitic, and have a nonphotosynthetichloroplast. They were once thought to be related to the helicosproidia, but it is now known that the helicosproida are green algae rather than chromalveolates. The apicomplexans include *Plasmodium*, the malaria parasite. Many apicomplexans keep a vestigial red algal derived chloroplast called an apicoplast, which they inherited from their ancestors. Other apicomplexans like *Cryptosporidium* have lost the chloroplast completely. Apicomplexans store their energy in amylopectin starch granules that are located in their cytoplasm, even though they are nonphotosynthetic.

Apicoplasts have lost all photosynthetifunction, and contain no photosynthetipigments or true thylakoids. They are bounded by four membranes, but the membranes are not connected to the endoplasmireticulum. The fact that apicomplexans still keep their nonphotosynthetichloroplast around demonstrates how the chloroplast carries out important functions other than photosynthesis. Plant chloroplasts provide plant cells with many important things besides sugar, and apicoplasts are no different—they synthesize fatty acids, isopentenyl pyrophosphate, iron-sulfur clusters, and carry out part of the heme pathway. This makes the apicoplast an attractive target for drugs to cure apicomplexan-related diseases. The most important apicoplast function is isopentenyl pyrophosphate synthesis—in fact, apicomplexans die when something interferes with this apicoplast function, and when apicomplexans are grown in an isopentenyl pyrophosphate-rich medium, they dump the organelle.

Dinophytes

The dinoflagellates are yet another very large and diverse group of protists, around half of which are (at least partially) photosynthetic.

Most dinophyte chloroplasts are secondary red algal derived chloroplasts, like other chromalveolate chloroplasts. Many other dinophytes have lost the chloroplast (becoming the nonphotosynthetikind of dinoflagellate), or replaced it though *tertiary* endosymbiosis—the engulfment of another chromalveolate containing a red algal derived chloroplast. Others replaced their original chloroplast with a green algal derived one.

Most dinophyte chloroplasts contain at least the photosynthetipigments chlorophyll a, chlorophyll c_2, *beta*-carotene, and at least one dinophyte-unique xanthophyll (peridinin, dinoxanthin, or diadinoxanthin), giving many a golden-brown color. All dinophytes store starch in their cytoplasm, and most have chloroplasts with thylakoids arranged in stacks of three.

Peridinin-containing Dinophyte Chloroplast

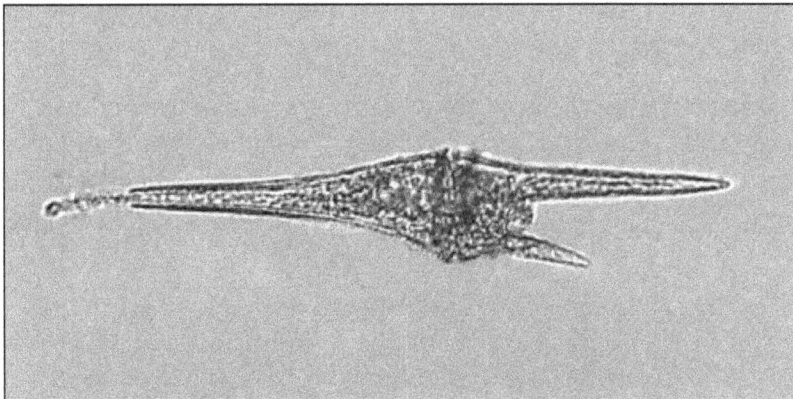

Ceratium furca, a peridinin-containing dinophyte

The most common dinophyte chloroplast is the peridinin-type chloroplast, characterized by the carotenoid pigment peridinin in their chloroplasts, along with chlorophyll a and chlorophyll c_2. Peridinin is not found in any other group of chloroplasts. The peridinin chloroplast is bounded by three membranes (occasionally two), having lost the red algal endosymbiont's original cell membrane. The outermost membrane is not connected to the endoplasmireticulum. They contain a pyrenoid, and have triplet-stacked thylakoids. Starch is found outside the chloroplast. An important feature of these chloroplasts is that their chloroplast DNA is highly reduced and fragmented into many small circles. Most of the genome has migrated to the nucleus, and only critical photosynthesis-related genes remain in the chloroplast.

The peridinin chloroplast is thought to be the dinophytes' "original" chloroplast, which has been lost, reduced, replaced, or has company in several other dinophyte lineages.

Fucoxanthin-containing Dinophyte Chloroplasts (Haptophyte Endosymbionts)

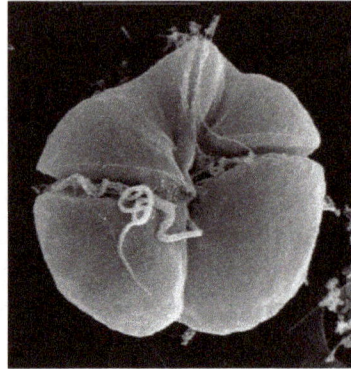

Karenia brevis is a fucoxanthin-containing dynophyte responsible for algal blooms called "red tides".

The fucoxanthin dinophyte lineages (including *Karlodinium* and *Karenia*) lost their original red algal derived chloroplast, and replaced it with a new chloroplast derived from a haptophyte endosymbiont. *Karlodinium* and *Karenia* probably took up different heterokontophytes. Because the haptophyte chloroplast has four membranes, tertiary endosymbiosis would be expected to create a six membraned chloroplast, adding the haptophyte's cell membrane and the dinophyte's phagosomal vacuole. However, the haptophyte was heavily reduced, stripped of a few membranes and its nucleus, leaving only its chloroplast (with its original double membrane), and possibly one or two additional membranes around it.

Fucoxanthin-containing chloroplasts are characterized by having the pigment fucoxanthin (actually 19'-hexanoyloxy-fucoxanthin and/or 19'-butanoyloxy-fucoxanthin) and no peridinin. Fucoxanthin is also found in haptophyte chloroplasts, providing evidence of ancestry.

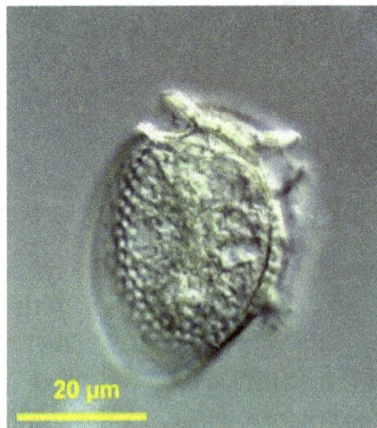

Dinophysis acuminata has chloroplasts taken from a cryptophyte.

Cryptophyte Derived Dinophyte Chloroplast

Members of the genus *Dinophysis* have a phycobilin-containing chloroplast taken from a cryptophyte. However, the cryptophyte is not an endosymbiont—only the chloroplast seems to have been taken, and the chloroplast has been stripped of its nucleomorph and outermost two membranes, leaving just a two-membraned chloroplast. Cryptophyte chloroplasts require their nucleomorph to maintain themselves, and *Dinophysis* species grown in cell culture alone cannot survive, so it is possible (but not confirmed) that the *Dinophysis* chloroplast is a kleptoplast—if so, *Dinophysis* chloroplasts wear out and *Dinophysis* species must continually engulf cryptophytes to obtain new chloroplasts to replace the old ones.

Diatom Derived Dinophyte Chloroplasts

Some dinophytes, like *Kryptoperidinium* and *Durinskia* have a diatom (heterokontophyte) derived chloroplast. These chloroplasts are bounded by up to *five* membranes, (depending on whether you count the entire diatom endosymbiont as the chloroplast, or just the red algal derived chloroplast inside it). The diatom endosymbiont has been reduced relatively little—it still retains its original mitochondria, and has endoplasmireticulum, ribosomes, a nucleus, and of course, red algal derived chloroplasts—practically a complete cell, all inside the host's endoplasmireticulum lumen. However the diatom endosymbiont can't store its own food—its starch is found in granules in the dinophyte host's cytoplasm instead. The diatom endosymbiont's nucleus is present, but it probably can't be called a nucleomorph because it shows no sign of genome reduction, and might have even been *expanded*. Diatoms have been engulfed by dinoflagellates at least three times.

The diatom endosymbiont is bounded by a single membrane, inside it are chloroplasts with four membranes. Like the diatom endosymbiont's diatom ancestor, the chloroplasts have triplet thylakoids and pyrenoids.

In some of these genera, the diatom endosymbiont's chloroplasts aren't the only chloroplasts in the dinophyte. The original three-membraned peridinin chloroplast is still around, converted to an eyespot.

Prasinophyte (Green Algal) Derived Dinophyte Chloroplast

Lepidodinium viride and its close relatives are dinophytes that lost their original peridinin chloroplast and replaced it with a green algal derived chloroplast (more specifically, a prasinophyte). *Lepidodinium* is the only dinophyte that has a chloroplast that's not from the rhodoplast lineage. The chloroplast is surrounded by two membranes and has no nucleomorph—all the nucleomorph genes have been transferred to the dinophyte nucleus. The endosymbiotievent that led to this chloroplast was serial secondary endosymbiosis rather than tertiary endosymbiosis—the endosymbiont was a green alga containing a primary chloroplast (making a secondary chloroplast).

Chromatophores

While most chloroplasts originate from that first set of endosymbiotievents, *Paulinella chromatophora* is an exception that acquired a photosyntheticyanobacterial endosymbiont more recently. It is not clear whether that symbiont is closely related to the ancestral chloroplast of other eukaryotes. Being in the early stages of endosymbiosis, *Paulinella chromatophora* can offer some insights into how chloroplasts evolved. *Paulinella* cells contain one or two sausage shaped blue-green photosynthesizing structures called chromatophores, descended from the cyanobacterium *Synechococcus*. Chromatophores cannot survive outside their host. Chromatophore DNA is about a million base pairs long, containing around 850 protein encoding genes—far less than the three million base pair *Synechococcus* genome, but much larger than the approximately 150,000 base pair genome of the more assimilated chloroplast. Chromatophores have transferred much less of their DNA to the nucleus of their host. About 0.3–0.8% of the nuclear DNA in *Paulinella* is from the chromatophore, compared with 11–14% from the chloroplast in plants.

Kleptoplastidy

In some groups of mixotrophiprotists, like some dinoflagellates, chloroplasts are separated from a captured alga or diatom and used temporarily. These klepto chloroplasts may only have a lifetime of a few days and are then replaced.

Chloroplast DNA

Chloroplasts have their own DNA, often abbreviated as ctDNA, or cpDNA. It is also known as the plastome. Its existence was first proved in 1962, and first sequenced in 1986—when two Japanese research teams sequenced the chloroplast DNA of liverwort and tobacco. Since then, hundreds of chloroplast DNAs from various species have been sequenced, but they're mostly those of land plants and green algae—glaucophytes, red algae, and other algal groups are extremely underrepresented, potentially introducing some bias in views of "typical" chloroplast DNA structure and content.

Molecular Structure

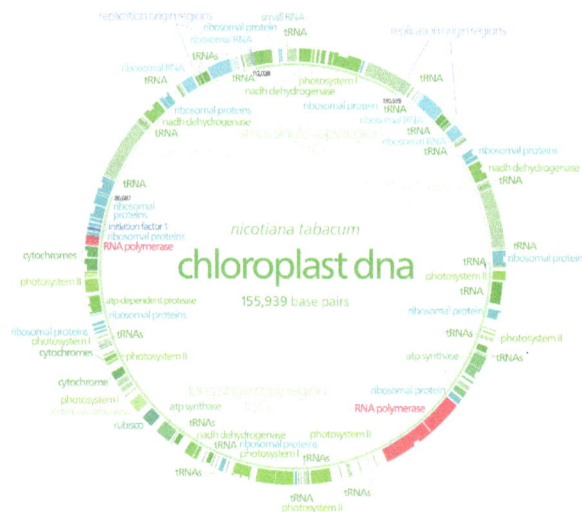

Chloroplast DNA Interactive gene map of chloroplast DNA from *Nicotiana tabacum*. Segments with labels on the inside reside on the B strand of DNA, segments with labels on the outside are on the A strand. Notches indicate introns.

With few exceptions, most chloroplasts have their entire chloroplast genome combined into a single large circular DNA molecule, typically 120,000–170,000 base pairs long. They can have a contour length of around 30–60 micrometers, and have a mass of about 80–130 million daltons.

While usually thought of as a circular molecule, there is some evidence that chloroplast DNA molecules more often take on a linear shape.

Inverted Repeats

Many chloroplast DNAs contain two *inverted repeats*, which separate a long single copy section (LSC) from a short single copy section (SSC). While a given pair of inverted repeats are rarely completely identical, they are always very similar to each other, apparently resulting from concerted evolution.

The inverted repeats vary wildly in length, ranging from 4,000 to 25,000 base pairs long each and containing as few as four or as many as over 150 genes. Inverted repeats in plants tend to be at the upper end of this range, each being 20,000–25,000 base pairs long.

The inverted repeat regions are highly conserved among land plants, and accumulate few mutations. Similar inverted repeats exist in the genomes of cyanobacteria and the other two chloroplast lineages (glaucophyta and rhodophyceae), suggesting that they predate the chloroplast, though some chloroplast DNAs have since lost or flipped the inverted repeats (making them direct repeats). It is possible that the inverted repeats help stabilize the rest of the chloroplast genome, as chloroplast DNAs which have lost some of the inverted repeat segments tend to get rearranged more.

Nucleoids

New chloroplasts may contain up to 100 copies of their DNA, though the number of chloroplast DNA copies decreases to about 15–20 as the chloroplasts age. They are usually packed into nucleoids, which can contain several identical chloroplast DNA rings. Many nucleoids can be found in each chloroplast. In primitive red algae, the chloroplast DNA nucleoids are clustered in the center of the chloroplast, while in green plants and green algae, the nucleoids are dispersed throughout the stroma.

Though chloroplast DNA is not associated with true histones, in red algae, similar proteins that tightly pack each chloroplast DNA ring into a nucleoid have been found.

DNA Replication

The Leading Model of cpDNA Replication

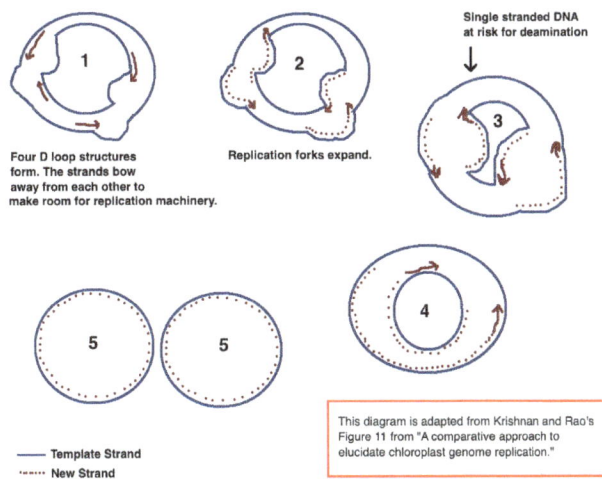

Chloroplast DNA replication via multiple D loop mechanisms. Adapted from Krishnan NM, Rao BJ's paper "A comparative approach to elucidate chloroplast genome replication."

The mechanism for chloroplast DNA (cpDNA) replication has not been conclusively determined, but two main models have been proposed. Scientists have attempted to observe chloroplast replication via electron microscopy since the 1970s. The results of the microscopy experiments led to the idea that chloroplast DNA replicates using a double displacement loop (D-loop). As the D-loop moves through the circular DNA, it adopts a theta intermediary form, also known as a Cairns replication intermediate, and completes replication with a rolling circle mechanism. Transcription starts at specifipoints of origin. Multiple replication forks open up, allowing replication machinery to transcribe the DNA. As replication continues, the forks grow and eventually converge. The new cpDNA structures separate, creating daughter cpDNA chromosomes.

In addition to the early microscopy experiments, this model is also supported by the amounts of deamination seen in cpDNA. Deamination occurs when an amino group is lost and is a mutation that often results in base changes. When adenine is deaminated, it becomes hypoxanthine. Hypoxanthine can bind to cytosine, and when the Xbase pair is replicated, it becomes a G(thus, an A--> G base change).

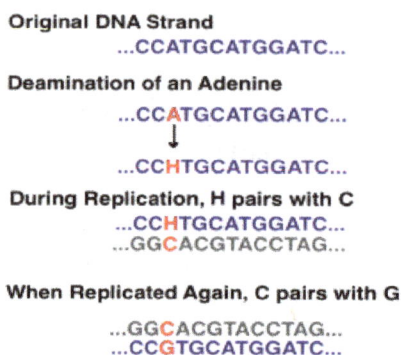

Original DNA Strand
...CCATGCATGGATC...

Deamination of an Adenine
...CCATGCATGGATC...
↓
...CCHTGCATGGATC...

During Replication, H pairs with C
...CCHTGCATGGATC...
...GGCACGTACCTAG...

When Replicated Again, C pairs with G
...GGCACGTACCTAG...
...CCGTGCATGGATC...

Over time, base changes in the DNA sequence can arise from deamination mutations. When adenine is deaminated, it becomes hypoxanthine, which can pair with cytosine. During replication, the cytosine will pair with guanine, causing an A --> G base change.

Deamination

In cpDNA, there are several A -->G deamination gradients. DNA becomes susceptible to deamination events when it is single stranded. When replication forks form, the strand not being copied is single stranded, and thus at risk for A -->G deamination. Therefore, gradients in deamination indicate that replication forks were most likely present and the direction that they initially opened (the highest gradient is most likely nearest the start site because it was single stranded for the longest amount of time). This mechanism is still the leading theory today; however, a second theory suggests that most cpDNA is actually linear and replicates through homologous recombination. It further contends that only a minority of the genetimaterial is kept in circular chromosomes while the rest is in branched, linear, or other complex structures.

Alternative Model of Replication

One of competing model for cpDNA replication asserts that most cpDNA is linear and participates in homologous recombination and replication structures similar to bacteriophage T4. It has been established that some plants have linear cpDNA, such as maize, and that more species still contain complex structures that scientists do not yet understand. When the original experiments on cpDNA were performed, scientists did notice linear structures; however, they attributed these linear forms to broken circles. If the branched and complex structures seen in cpDNA experiments are real and not artifacts of concatenated circular DNA or broken circles, then a D-loop mechanism of replication is insufficient to explain how those structures would replicate. At the same time, homologous recombination does not expand the multiple A --> G gradients seen in plastomes. Because of the failure to explain the deamination gradient as well as the numerous plant species that have been shown to have circular cpDNA, the predominant theory continues to hold that most cpDNA is circular and most likely replicates via a D loop mechanism.

Gene Content and Protein Synthesis

The chloroplast genome most commonly includes around 100 genes that code for a variety of things, mostly to do with the protein pipeline and photosynthesis. As in prokaryotes, genes in chloroplast DNA are organized into operons. Interestingly though, unlike prokaryotiDNA molecules, chloroplast DNA molecules contain introns (plant mitochondrial DNAs do too, but not human mtDNAs).

Among land plants, the contents of the chloroplast genome are fairly similar.

Chloroplast Genome Reduction and Gene Transfer

Over time, many parts of the chloroplast genome were transferred to the nuclear genome of the host, a process called *endosymbiotigene transfer*. As a result, the chloroplast genome is heavily reduced compared to that of free-living cyanobacteria. Chloroplasts may contain 60–100 genes whereas cyanobacteria often have more than 1500 genes in their genome. Recently, a plastid without a genome was found, demonstrating chloroplasts can lose their genome during endosymbiotigene transfer process.

Endosymbiotigene transfer is how we know about the lost chloroplasts in many chromalveolate lineages. Even if a chloroplast is eventually lost, the genes it donated to the former host's nucleus persist, providing evidence for the lost chloroplast's existence. For example, while diatoms (a heterokontophyte) now have a red algal derived chloroplast, the presence of many green algal genes in the diatom nucleus provide evidence that the diatom ancestor (probably the ancestor of all chromalveolates too) had a green algal derived chloroplast at some point, which was subsequently replaced by the red chloroplast.

In land plants, some 11–14% of the DNA in their nuclei can be traced back to the chloroplast, up to 18% in *Arabidopsis*, corresponding to about 4,500 protein-coding genes. There have been a few recent transfers of genes from the chloroplast DNA to the nuclear genome in land plants.

Of the approximately 3000 proteins found in chloroplasts, some 95% of them are encoded by nuclear genes. Many of the chloroplast's protein complexes consist of subunits from both the chloroplast genome and the host's nuclear genome. As a result, protein synthesis must be coordinated between the chloroplast and the nucleus. The chloroplast is mostly under nuclear control, though chloroplasts can also give out signals regulating gene expression in the nucleus, called *retrograde signaling*.

Protein Synthesis

Protein synthesis within chloroplasts relies on two RNA polymerases. One is coded by the chloroplast DNA, the other is of nuclear origin. The two RNA polymerases may recognize and bind to different kinds of promoters within the chloroplast genome. The ribosomes in chloroplasts are similar to bacterial ribosomes.

Protein Targeting and Import

Because so many chloroplast genes have been moved to the nucleus, many proteins that would originally have been translated in the chloroplast are now synthesized in the cytoplasm of the plant cell. These proteins must be directed back to the chloroplast, and imported through at least two chloroplast membranes.

Curiously, around half of the protein products of transferred genes aren't even targeted back to the chloroplast. Many became exaptations, taking on new functions like participating in cell division, protein routing, and even disease resistance. A few chloroplast genes found new homes in the mitochondrial genome—most became nonfunctional pseudogenes, though a few tRNA genes still work in the mitochondrion. Some transferred chloroplast DNA protein products get directed to the secretory pathway (though it should be noted that many secondary plastids are bounded by an outermost membrane derived from the host's cell membrane, and therefore topologically outside of the cell, because to reach the chloroplast from the cytosol, you have to cross the cell membrane, just like if you were headed for the extracellular space. In those cases, chloroplast-targeted proteins do initially travel along the secretory pathway).

Because the cell acquiring a chloroplast already had mitochondria (and peroxisomes, and a cell membrane for secretion), the new chloroplast host had to develop a unique protein targeting system to avoid having chloroplast proteins being sent to the wrong organelle, and the wrong proteins being sent to the chloroplast.

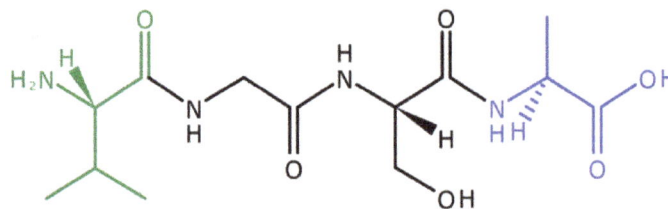

The two ends of a polypeptide are called the N-terminus, or *amino end*, and the C-terminus, or *carboxyl end*. This polypeptide has four amino acids linked together. At the left is the N-terminus, with its amino (H_2N) group in green. The blue C-terminus, with its carboxyl group (CO_2H) is at the right.

In most, but not all cases, nuclear-encoded chloroplast proteins are translated with a *cleavable transit peptide* that's added to the N-terminus of the protein precursor. Sometimes the transit sequence is found on the C-terminus of the protein, or within the functional part of the protein.

Transport Proteins and Membrane Translocons

After a chloroplast polypeptide is synthesized on a ribosome in the cytosol, an enzyme specifito chloroplast proteins phosphorylates, or adds a phosphate group to many (but not all) of them in their transit sequences. Phosphorylation helps many proteins bind the polypeptide, keeping it from folding prematurely. This is important because it prevents chloroplast proteins from assuming their active form and carrying out their chloroplast functions in the wrong place—the cytosol. At the same time, they have to keep just enough shape so that they can be recognized by the chloroplast. These proteins also help the polypeptide get imported into the chloroplast.

From here, chloroplast proteins bound for the stroma must pass through two protein complexes—the TOcomplex, or *translocon on the outer chloroplast membrane*, and the TItranslocon, or *translocon on the inner chloroplast membrane translocon*. Chloroplast polypeptide chains probably often travel through the two complexes at the same time, but the TIcomplex can also retrieve preproteins lost in the intermembrane space.

Structure

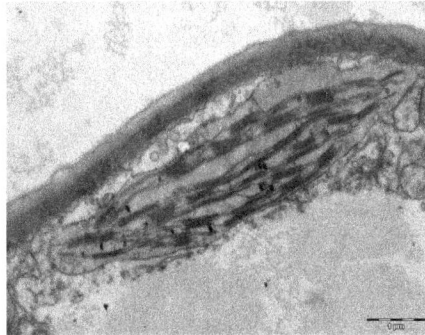

Transmission electron microscope image of a chloroplast. Grana of thylakoids and their connecting lamellae are clearly visible.

In land plants, chloroplasts are generally lens-shaped, 5–8 μm in diameter and 1–3 μm thick. Greater diversity in chloroplast shapes exists among the algae, which often contain a single chloroplast that can be shaped like a net (e.g., *Oedogonium*), a cup (e.g., *Chlamydomonas*), a ribbon-like spiral around the edges of the cell (e.g., *Spirogyra*), or slightly twisted bands at the cell edges (e.g., *Sirogonium*). Some algae have two chloroplasts in each cell; they are star-shaped in *Zygnema*, or may follow the shape of half the cell in order Desmidiales. In some algae, the chloroplast takes up most of the cell, with pockets for the nucleus and other organelles (for example some species of *Chlorella* have a cup-shaped chloroplast that occupies much of the cell).

All chloroplasts have at least three membrane systems—the outer chloroplast membrane, the inner chloroplast membrane, and the thylakoid system. Chloroplasts that are the product of secondary endosymbiosis may have additional membranes surrounding these three. Inside the outer and inner chloroplast membranes is the chloroplast stroma, a semi-gel-like fluid that makes up much of a chloroplast's volume, and in which the thylakoid system floats.

There are some common misconceptions about the outer and inner chloroplast membranes. The fact that chloroplasts are surrounded by a double membrane is often cited as evidence that they are the descendants of endosymbioticcyanobacteria. This is often interpreted as meaning the outer chloroplast membrane is the product of the host's cell membrane infolding to form a vesicle to surround the ancestral cyanobacterium—which is not true—both chloroplast membranes are homologous to the cyanobacterium's original double membranes.

The chloroplast double membrane is also often compared to the mitochondrial double membrane. This is not a valid comparison—the inner mitochondria membrane is used to run proton pumps and carry out oxidative phosphorylation across to generate ATP energy. The only chloroplast structure that can considered analogous to it is the internal thylakoid system. Even so, in terms of "in-out", the direction of chloroplast H⁺ ion flow is in the opposite direction compared to oxidative phosphorylation in mitochondria. In addition, in terms of function, the inner chloroplast membrane, which regulates metabolite passage and synthesizes some materials, has no counterpart in the mitochondrion.

Outer Chloroplast Membrane

The outer chloroplast membrane is a semi-porous membrane that small molecules and ions can easily diffuse across. However, it is not permeable to larger proteins, so chloroplast polypeptides being synthesized in the cell cytoplasm must be transported across the outer chloroplast membrane by the TOcomplex, or *translocon on the outer chloroplast* membrane.

The chloroplast membranes sometimes protrude out into the cytoplasm, forming a stromule, or stroma-containing tubule. Stromules are very rare in chloroplasts, and are much more common in other plastids like chromoplasts and amyloplasts in petals and roots, respectively. They may exist to increase the chloroplast's surface area for cross-membrane transport, because they are often branched and tangled with the endoplasmireticulum. When they were first observed in 1962, some plant biologists dismissed the structures as artifactual, claiming that stromules were just oddly shaped chloroplasts with constricted regions or dividing chloroplasts. However, there is a growing body of evidence that stromules are functional, integral features of plant cell plastids, not merely artifacts.

Intermembrane Space and Peptidoglycan Wall

Instead of an intermembrane space, glaucophyte algae have a peptidoglycan wall between their inner and outer chloroplast membranes.

Usually, a thin intermembrane space about 10–20 nanometers thick exists between the outer and inner chloroplast membranes.

Glaucophyte algal chloroplasts have a peptidoglycan layer between the chloroplast membranes. It corresponds to the peptidoglycan cell wall of their cyanobacterial ancestors, which is located between their two cell membranes. These chloroplasts are called muroplasts. Other chloroplasts have lost the cyanobacterial wall, leaving an intermembrane space between the two chloroplast envelope membranes.

Inner Chloroplast Membrane

The inner chloroplast membrane borders the stroma and regulates passage of materials in and out of the chloroplast. After passing through the TOcomplex in the outer chloroplast membrane, polypeptides must pass through the TIcomplex *(translocon on the inner chloroplast membrane)* which is located in the inner chloroplast membrane.

In addition to regulating the passage of materials, the inner chloroplast membrane is where fatty acids, lipids, and carotenoids are synthesized.

Peripheral Reticulum

Some chloroplasts contain a structure called the chloroplast peripheral reticulum. It is often found in the chloroplasts of C_4 plants, though it has also been found in some C_3 angiosperms, and even some gymnosperms. The chloroplast peripheral reticulum consists of a maze of membranous tubes and vesicles continuous with the inner chloroplast membrane that extends into the internal stromal fluid of the chloroplast. Its purpose is thought to be to increase the chloroplast's surface area for cross-membrane transport between its stroma and the cell cytoplasm. The small vesicles sometimes observed may serve as transport vesicles to shuttle stuff between the thylakoids and intermembrane space.

Stroma

The protein-rich, alkaline, aqueous fluid within the inner chloroplast membrane and outside of the thylakoid space is called the stroma, which corresponds to the cytosol of the original cyanobacterium. Nucleoids of chloroplast DNA, chloroplast ribosomes, the thylakoid system with plastoglobuli, starch granules, and many proteins can be found floating around in it. The Calvin cycle, which fixes CO_2 into sugar takes place in the stroma.

Chloroplast Ribosomes

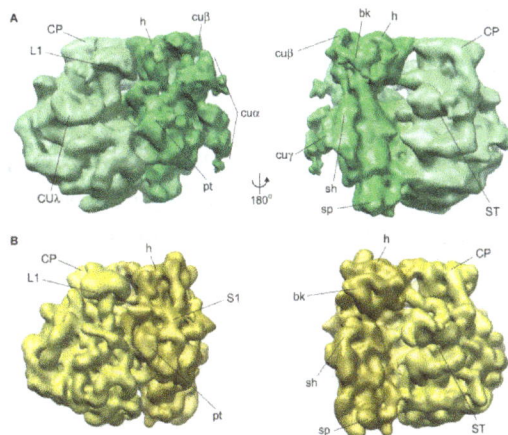

Chloroplast ribosomes Comparison of a chloroplast ribosome (green) and a bacterial ribosome (yellow). Important features common to both ribosomes and chloroplast-unique features are labeled.

Chloroplasts have their own ribosomes, which they use to synthesize a small fraction of their proteins. Chloroplast ribosomes are about two-thirds the size of cytoplasmiribosomes (around 17 nm vs 25 nm). They take mRNAs transcribed from the chloroplast DNA and translate them into protein. While similar to bacterial ribosomes, chloroplast translation is more complex than in bacteria, so chloroplast ribosomes include some chloroplast-unique features. Small subunit ribosomal RNAs in several Chlorophyta and euglenid chloroplasts lack motifs for shine-dalgarno sequence recognition, which is considered essential for translation initiation in most chloroplasts and prokaryotes. Such loss is also rarely observed in other plastids and prokaryotes.

Plastoglobuli

Plastoglobuli (singular *plastoglobulus*, sometimes spelled *plastoglobule(s)*), are spherical bubbles of lipids and proteins about 45–60 nanometers across. They are surrounded by a lipid monolayer. Plastoglobuli are found in all chloroplasts, but become more common when the chloroplast is under oxidative stress, or when it ages and transitions into a gerontoplast. Plastoglobuli also exhibit a greater size variation under these conditions. They are also common in etioplasts, but decrease in number as the etioplasts mature into chloroplasts.

Plastoglubuli contain both structural proteins and enzymes involved in lipid synthesis and metabolism. They contain many types of lipids including plastoquinone, vitamin E, carotenoids and chlorophylls.

Plastoglobuli were once thought to be free-floating in the stroma, but it is now thought that they are permanently attached either to a thylakoid or to another plastoglobulus attached to a thylakoid, a configuration that allows a plastoglobulus to exchange its contents with the thylakoid network. In normal green chloroplasts, the vast majority of plastoglobuli occur singularly, attached directly to their parent thylakoid. In old or stressed chloroplasts, plastoglobuli tend to occur in linked groups or chains, still always anchored to a thylakoid.

Plastoglobuli form when a bubble appears between the layers of the lipid bilayer of the thylakoid membrane, or bud from existing plastoglubuli—though they never detach and float off into the stroma. Practically all plastoglobuli form on or near the highly curved edges of the thylakoid disks or sheets. They are also more common on stromal thylakoids than on granal ones.

Starch Granules

Starch granules are very common in chloroplasts, typically taking up 15% of the organelle's volume, though in some other plastids like amyloplasts, they can be big enough to distort the shape of the organelle. Starch granules are simply accumulations of starch in the stroma, and are not bounded by a membrane.

Starch granules appear and grow throughout the day, as the chloroplast synthesizes sugars, and are consumed at night to fuel respiration and continue sugar export into the phloem, though in mature chloroplasts, it is rare for a starch granule to be completely consumed or for a new granule to accumulate.

Starch granules vary in composition and location across different chloroplast lineages. In red algae, starch granules are found in the cytoplasm rather than in the chloroplast. In C_4 plants, mesophyll chloroplasts, which do not synthesize sugars, lack starch granules.

Rubisco

The chloroplast stroma contains many proteins, though the most common and important is Rubisco, which is probably also the most abundant protein on the planet. Rubisco is the enzyme that fixes CO_2 into sugar molecules. In C_3 plants, rubisco is abundant in all chloroplasts, though in C_4 plants, it is confined to the bundle sheath chloroplasts, where the Calvin cycle is carried out in C_4 plants.

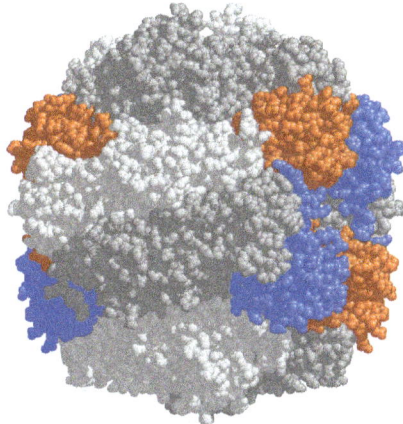

Rubisco, shown here in a space-filling model, is the main enzyme responsible for carbon fixation in chloroplasts.

Pyrenoids

The chloroplasts of some hornworts and algae contain structures called pyrenoids. They are not found in higher plants. Pyrenoids are roughly spherical and highly refractive bodies which are a site of starch accumulation in plants that contain them. They consist of a matrix opaque to electrons, surrounded by two hemispherical starch plates. The starch is accumulated as the pyrenoids mature. In algae with carbon concentrating mechanisms, the enzyme rubisco is found in the pyrenoids. Starch can also accumulate around the pyrenoids when CO_2 is scarce. Pyrenoids can divide to form new pyrenoids, or be produced "de novo".

Thylakoid System

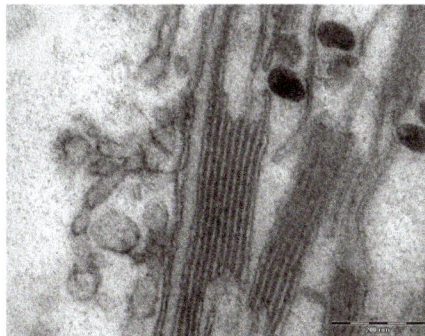

Transmission electron microscope image of some thylakoids arranged in grana stacks and lamellæ. Plastoglobuli (dark blobs) are also present.

Suspended within the chloroplast stroma is the thylakoid system, a highly dynamiccollection of membranous sacks called thylakoids where chlorophyll is found and the light reactions of photosynthesis happen. In most vascular plant chloroplasts, the thylakoids are arranged in stacks called grana, though in certain C_4 plant chloroplasts and some algal chloroplasts, the thylakoids are free floating.

Granal Structure

Using a light microscope, it is just barely possible to see tiny green granules—which were named grana. With electron microscopy, it became possible to see the thylakoid system in more detail, revealing it to consist of stacks of flat thylakoids which made up the grana, and long interconnecting stromal thylakoids which linked different grana. In the transmission electron microscope, thylakoid membranes appear as alternating light-and-dark bands, 8.5 nanometers thick.

For a long time, the three-dimensional structure of the thylakoid system has been unknown or disputed. One model has the granum as a stack of thylakoids linked by helical stromal thylakoids; the other has the granum as a single folded thylakoid connected in a "hub and spoke" way to other grana by stromal thylakoids. While the thylakoid system is still commonly depicted according to the folded thylakoid model, it was determined in 2011 that the stacked and helical thylakoids model is correct.

In the helical thylakoid model, grana consist of a stack of flattened circular granal thylakoids that resemble pancakes. Each granum can contain anywhere from two to a hundred thylakoids, though grana with 10–20 thylakoids are most common. Wrapped around the grana are helicoid stromal thylakoids, also known as frets or lamellar thylakoids. The helices ascend at an angle of 20–25°, connecting to each granal thylakoid at a bridge-like slit junction. The helicoids may extend as large sheets that link multiple grana, or narrow to tube-like bridges between grana. While different parts of the thylakoid system contain different membrane proteins, the thylakoid membranes are continuous and the thylakoid space they enclose form a single continuous labyrinth.

Thylakoids

Thylakoids (sometimes spelled *thylakoïds*), are small interconnected sacks which contain the membranes that the light reactions of photosynthesis take place on. The word *thylakoid* comes from the Greek word *thylakos* which means "sack".

Embedded in the thylakoid membranes are important protein complexes which carry out the light reactions of photosynthesis. Photosystem II and photosystem I contain light-harvesting complexes with chlorophyll and carotenoids that absorb light energy and use it to energize electrons. Molecules in the thylakoid membrane use the energized electrons to pump hydrogen ions into the thylakoid space, decreasing the pH and turning it acidic. ATP synthase is a large protein complex that harnesses the concentration gradient of the hydrogen ions in the thylakoid space to generate ATP energy as the hydrogen ions flow back out into the stroma—much like a dam turbine.

There are two types of thylakoids—granal thylakoids, which are arranged in grana, and stromal thylakoids, which are in contact with the stroma. Granal thylakoids are pancake-shaped circular disks about 300–600 nanometers in diameter. Stromal thylakoids are helicoid sheets that spiral around grana. The flat tops and bottoms of granal thylakoids contain only the relatively flat photosystem II protein complex. This allows them to stack tightly, forming grana with many layers of tightly appressed membrane, called granal membrane, increasing stability and surface area for light capture.

In contrast, photosystem I and ATP synthase are large protein complexes which jut out into the stroma. They can't fit in the appressed granal membranes, and so are found in the stromal thylakoid membrane—the edges of the granal thylakoid disks and the stromal thylakoids. These large protein complexes may act as spacers between the sheets of stromal thylakoids.

The number of thylakoids and the total thylakoid area of a chloroplast is influenced by light exposure. Shaded chloroplasts contain larger and more grana with more thylakoid membrane area than chloroplasts exposed to bright light, which have smaller and fewer grana and less thylakoid area. Thylakoid extent can change within minutes of light exposure or removal.

Pigments and Chloroplast Colors

Inside the photosystems embedded in chloroplast thylakoid membranes are various photosynthetipigments, which absorb and transfer light energy. The types of pigments found are different in various groups of chloroplasts, and are responsible for a wide variety of chloroplast colorations.

Paper chroma-tography of some spinach leaf extract shows the various pigments present in their chloroplasts.

Chlorophylls

Chlorophyll a is found in all chloroplasts, as well as their cyanobacterial ancestors. Chlorophyll a is a blue-green pigment partially responsible for giving most cyanobacteria and chloroplasts their color. Other forms of chlorophyll exist, such as the accessory pigments chlorophyll b, chlorophyll c, chlorophyll d, and chlorophyll f.

Chlorophyll b is an olive green pigment found only in the chloroplasts of plants, green algae, any secondary chloroplasts obtained through the secondary endosymbiosis of a green alga, and a few cyanobacteria. It is the chlorophylls a and b together that make most plant and green algal chloroplasts green.

Chlorophyll is mainly found in secondary endosymbiotichloroplasts that originated from a red alga, although it is not found in chloroplasts of red algae themselves. Chlorophyll is also found in some green algae and cyanobacteria.

Chlorophylls d and f are pigments found only in some cyanobacteria.

Carotenoids

Delesseria sanguinea, a red alga, has chloroplasts that contain red pigments like phycoerytherin that mask their blue-green chlorophyll *a*.

In addition to chlorophylls, another group of yellow–orange pigments called carotenoids are also found in the photosystems. There are about thirty photosyntheticarotenoids. They help transfer and dissipate excess energy, and their bright colors sometimes override the chlorophyll green, like during the fall, when the leaves of some land plants change color. β-carotene is a bright red-orange carotenoid found in nearly all chloroplasts, like chlorophyll *a*. Xanthophylls, especially the orange-red zeaxanthin, are also common. Many other forms of carotenoids exist that are only found in certain groups of chloroplasts.

Phycobilins

Phycobilins are a third group of pigments found in cyanobacteria, and glaucophyte, red algal, and cryptophyte chloroplasts. Phycobilins come in all colors, though phycoerytherin is one of the pigments that makes many red algae red. Phycobilins often organize into relatively large protein complexes about 40 nanometers across called phycobilisomes. Like photosystem I and ATP synthase, phycobilisomes jut into the stroma, preventing thylakoid stacking in red algal chloroplasts. Cryptophyte chloroplasts and some cyanobacteria don't have their phycobilin pigments organized into phycobilisomes, and keep them in their thylakoid space instead.

Specialized Chloroplasts in C$_4$ Plants

To fix carbon dioxide into sugar molecules in the process of photosynthesis, chloroplasts use an enzyme called rubisco. Rubisco has a problem—it has trouble distinguishing between carbon dioxide and oxygen, so at high oxygen concentrations, rubisco starts accidentally adding oxygen to sugar precursors. This has the end result of ATP energy being wasted and CO_2 being released, all with no sugar being produced. This is a big problem, since O_2 is produced by the initial light reactions of photosynthesis, causing issues down the line in the Calvin cycle which uses rubisco.

C$_4$ plants evolved a way to solve this—by spatially separating the light reactions and the Calvin cycle. The light reactions, which store light energy in ATP and NADPH, are done in the mesophyll cells of a C$_4$ leaf. The Calvin cycle, which uses the stored energy to make sugar using rubisco, is done in the bundle sheath cells, a layer of cells surrounding a vein in a leaf.

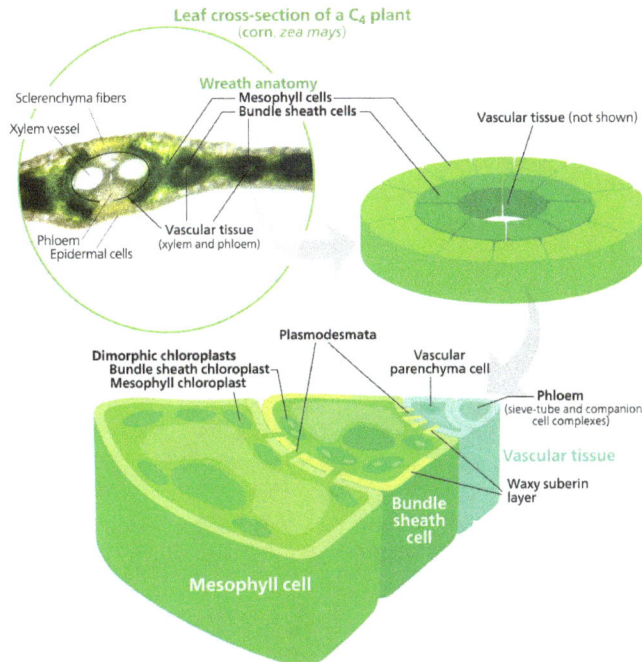

Many C$_4$ plants have their mesophyll cells and bundle sheath cells arranged radially around their leaf veins. The two types of cells contain different types of chloroplasts specialized for a particular part of photosynthesis.

As a result, chloroplasts in C$_4$ mesophyll cells and bundle sheath cells are specialized for each stage of photosynthesis. In mesophyll cells, chloroplasts are specialized for the light reactions, so they lack rubisco, and have normal grana and thylakoids, which they use to make ATP and NADPH, as well as oxygen. They store CO$_2$ in a four-carbon compound, which is why the process is called C$_4$ *photosynthesis*. The four-carbon compound is then transported to the bundle sheath chloroplasts, where it drops off CO$_2$ and returns to the mesophyll. Bundle sheath chloroplasts do not carry out the light reactions, preventing oxygen from building up in them and disrupting rubisco activity. Because of this, they lack thylakoids organized into grana stacks—though bundle sheath chloroplasts still have free-floating thylakoids in the stroma where they still carry out cyclielectron flow, a light-driven method of synthesizing ATP to power the Calvin cycle without generating oxygen. They lack photosystem II, and only have photosystem I—the only protein complex needed for cyclielectron flow. Because the job of bundle sheath chloroplasts is to carry out the Calvin cycle and make sugar, they often contain large starch grains.

Both types of chloroplast contain large amounts of chloroplast peripheral reticulum, which they use to get more surface area to transport stuff in and out of them. Mesophyll chloroplasts have a little more peripheral reticulum than bundle sheath chloroplasts.

Location

Distribution in a Plant

Not all cells in a multicellular plant contain chloroplasts. All green parts of a plant contain chloroplasts—the chloroplasts, or more specifically, the chlorophyll in them are what make the photosynthetiparts of a plant green. The plant cells which contain chloroplasts are usually parenchyma cells, though chloroplasts can also be found in collenchyma tissue. A plant cell which contains chloroplasts is known as a chlorenchyma cell. A typical chlorenchyma cell of a land plant contains about 10 to 100 chloroplasts.

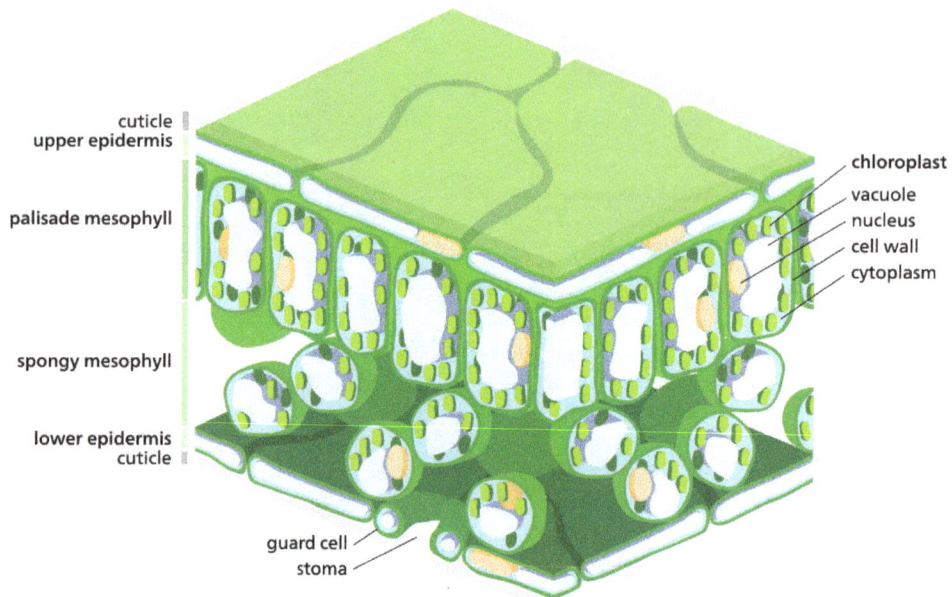

A cross section of a leaf, showing chloroplasts in its mesophyll cells. Stomal guard cells also have chloroplasts, though much fewer than mesophyll cells.

In some plants such as cacti, chloroplasts are found in the stems, though in most plants, chloroplasts are concentrated in the leaves. One square millimeter of leaf tissue can contain half a million chloroplasts. Within a leaf, chloroplasts are mainly found in the mesophyll layers of a leaf, and the guard cells of stomata. Palisade mesophyll cells can contain 30–70 chloroplasts per cell, while stomatal guard cells contain only around 8–15 per cell, as well as much less chlorophyll. Chloroplasts can also be found in the bundle sheath cells of a leaf, especially in C_4 plants, which carry out the Calvin cycle in their bundle sheath cells. They are often absent from the epidermis of a leaf.

Cellular Location

Chloroplast Movement

The chloroplasts of plant and algal cells can orient themselves to best suit the available light. In low-light conditions, they will spread out in a sheet—maximizing the surface area to absorb light. Under intense light, they will seek shelter by aligning in vertical columns along the plant cell's cell wall or turning sideways so that light strikes them edge-on. This reduces exposure and protects them from photooxidative damage. This ability to distribute chloroplasts so that they can take shelter behind each other or spread out may be the reason why land plants evolved to have many small chloroplasts instead of a few big ones. Chloroplast movement is considered one of the most closely regulated stimulus-response systems that can be found in plants. Mitochondria have also been observed to follow chloroplasts as they move.

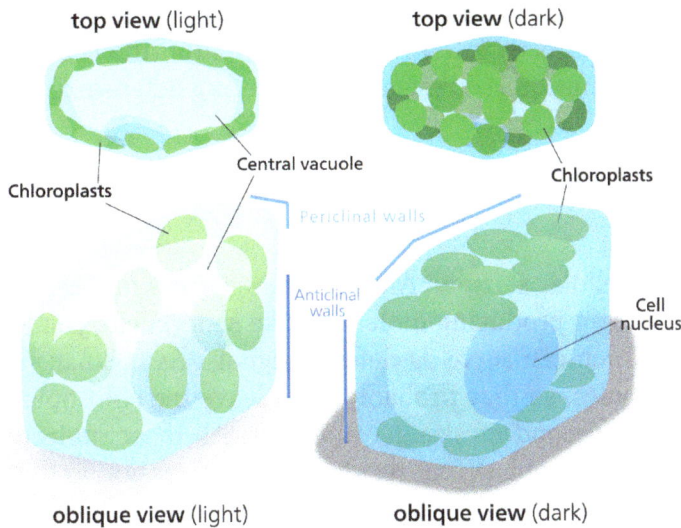

top view (light)   top view (dark)

Chloroplasts   Central vacuole   Chloroplasts

Periclinal walls

Anticlinal walls   Cell nucleus

oblique view (light)   oblique view (dark)

When chloroplasts are exposed to direct sunlight, they stack along the anticlinal cell walls to minimize exposure. In the dark they spread out in sheets along the periclinal walls to maximize light absorption.

In higher plants, chloroplast movement is run by phototropins, blue light photoreceptors also responsible for plant phototropism. In some algae, mosses, ferns, and flowering plants, chloroplast movement is influenced by red light in addition to blue light, though very long red wavelengths inhibit movement rather than speeding it up. Blue light generally causes chloroplasts to seek shelter, while red light draws them out to maximize light absorption.

Studies of *Vallisneria gigantea*, an aquatiflowering plant, have shown that chloroplasts can get moving within five minutes of light exposure, though they don't initially show any net directionality. They may move along microfilament tracks, and the fact that the microfilament mesh changes shape to form a honeycomb structure surrounding the chloroplasts after they have moved suggests that microfilaments may help to anchor chloroplasts in place.

Function and Chemistry

Guard Cell Chloroplasts

Unlike most epidermal cells, the guard cells of plant stomata contain relatively well developed chloroplasts. However, exactly what they do is controversial.

Plant Innate Immunity

Plants lack specialized immune cells—all plant cells participate in the plant immune response. Chloroplasts, along with the nucleus, cell membrane, and endoplasmireticulum, are key players in pathogen defense. Due to its role in a plant cell's immune response, pathogens frequently target the chloroplast.

Plants have two main immune responses—the hypersensitive response, in which infected cells seal themselves off and undergo programmed cell death, and systemiacquired resistance, where infected cells release signals warning the rest of the plant of a pathogen's presence. Chloroplasts stimulate both responses by purposely damaging their photosynthetisystem, producing reactive oxygen species. High levels of reactive oxygen species will cause the hypersensitive response. The reactive oxygen species also directly kill any pathogens within the cell. Lower levels of reactive oxygen species initiate systemiacquired resistance, triggering defense-molecule production in the rest of the plant.

In some plants, chloroplasts are known to move closer to the infection site and the nucleus during an infection.

Chloroplasts can serve as cellular sensors. After detecting stress in a cell, which might be due to a pathogen, chloroplasts begin producing molecules like salicyliacid, jasmoniacid, nitrioxide and reactive oxygen species which can serve as defense-signals. As cellular signals, reactive oxygen species are unstable molecules, so they probably don't leave the chloroplast, but instead pass on their signal to an unknown second messenger molecule. All these molecules initiate retrograde signaling—signals from the chloroplast that regulate gene expression in the nucleus.

In addition to defense signaling, chloroplasts, with the help of the peroxisomes, help synthesize an important defense molecule, jasmonate. Chloroplasts synthesize all the fatty acids in a plant cell—linoleiacid, a fatty acid, is a precursor to jasmonate.

Photosynthesis

One of the main functions of the chloroplast is its role in photosynthesis, the process by which light is transformed into chemical energy, to subsequently produce food in the form of sugars. Water (H_2O) and carbon dioxide (CO_2) are used in photosynthesis, and sugar and oxygen (O_2) is made, using light energy. Photosynthesis is divided into two stages—the light reactions, where water is split to produce oxygen, and the dark reactions, or Calvin cycle, which builds sugar molecules from carbon dioxide. The two phases are linked by the energy carriers adenosine triphosphate (ATP) and nicotinamide adenine dinucleotide phosphate ($NADP^+$).

Light Reactions

The light reactions of photosynthesis take place across the thylakoid membranes.

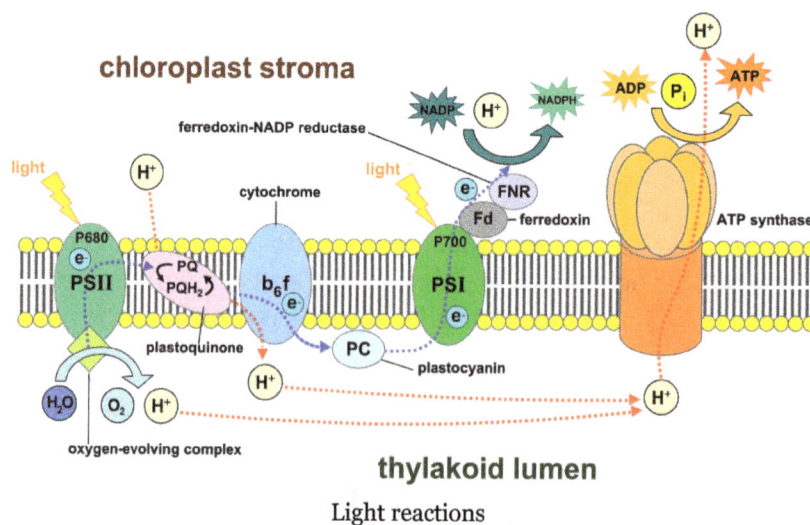

Light reactions

The light reactions take place on the thylakoid membranes. They take light energy and store it in NADPH, a form of $NADP^+$, and ATP to fuel the dark reactions.

Energy Carriers

ATP is the phosphorylated version of adenosine diphosphate (ADP), which stores energy in a cell and powers most cellular activities. ATP is the energized form, while ADP is the (partially) depleted form. $NADP^+$ is an electron carrier which ferries high energy electrons. In the light reactions, it gets reduced, meaning it picks up electrons, becoming NADPH.

Photophosphorylation

Like mitochondria, chloroplasts use the potential energy stored in an H^+, or hydrogen ion gradient to generate ATP energy. The two photosystems capture light energy to energize electrons taken from water, and release them down an electron transport chain. The molecules between the photosystems harness the electrons' energy to pump hydrogen ions into the thylakoid space, creating a concentration gradient, with more hydrogen ions (up to a thousand times as many) inside the thylakoid system than in the stroma. The hydrogen ions in the thylakoid space then diffuse back down their concentration gradient, flowing back out into the stroma through ATP synthase. ATP synthase uses the energy from the flowing hydrogen ions to phosphorylate adenosine diphosphate into adenosine triphosphate, or ATP. Because chloroplast ATP synthase projects out into the stroma, the ATP is synthesized there, in position to be used in the dark reactions.

$NADP^+$ Reduction

Electrons are often removed from the electron transport chains to charge $NADP^+$ with electrons, reducing it to NADPH. Like ATP synthase, ferredoxin-$NADP^+$ reductase, the enzyme that reduces $NADP^+$, releases the NADPH it makes into the stroma, right where it is needed for the dark reactions.

Because $NADP^+$ reduction removes electrons from the electron transport chains, they must be replaced—the job of photosystem II, which splits water molecules (H_2O) to obtain the electrons from its hydrogen atoms.

Cycliphotophosphorylation

While photosystem II photolyzes water to obtain and energize new electrons, photosystem I simply reenergizes depleted electrons at the end of an electron transport chain. Normally, the reenergized electrons are taken by $NADP^+$, though sometimes they can flow back down more H^+-pumping electron transport chains to transport more hydrogen ions into the thylakoid space to generate more ATP. This is termed cycliphotophosphorylation because the electrons are recycled. Cycliphotophosphorylation is common in C_4 plants, which need more ATP than NADPH.

Dark Reactions

The Calvin cycle, also known as the dark reactions, is a series of biochemical reactions that fixes CO_2 into G3P sugar molecules and uses the energy and electrons from the ATP and NADPH made in the light reactions. The Calvin cycle takes place in the stroma of the chloroplast.

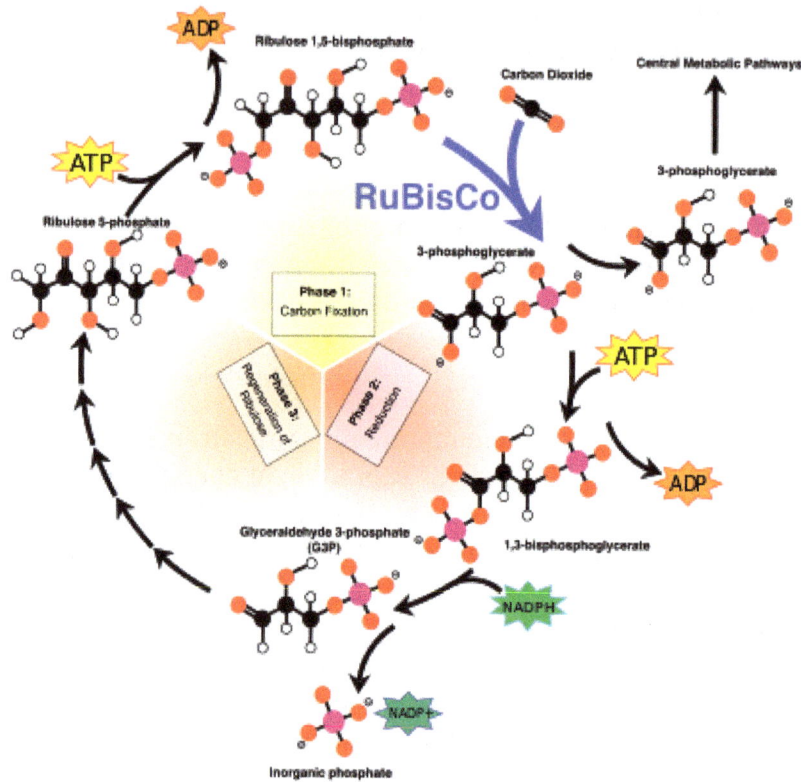

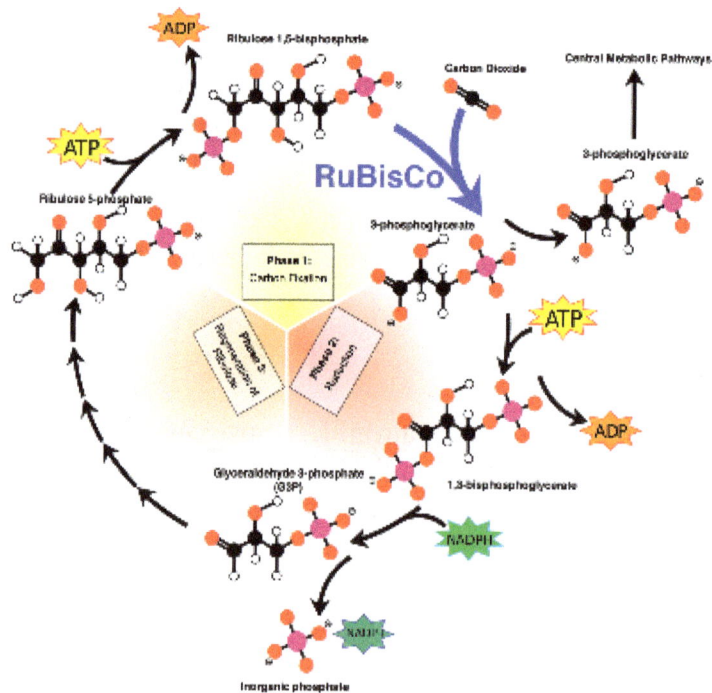

The Calvin cycle (*Interactive diagram*) The Calvin cycle incorporates carbon dioxide into sugar molecules.

While named *"the dark reactions"*, in most plants, they take place in the light, since the dark reactions are dependent on the products of the light reactions.

Carbon Fixation and G3P Synthesis

The Calvin cycle starts by using the enzyme Rubisco to fix CO_2 into five-carbon Ribulose bisphosphate (RuBP) molecules. The result is unstable six-carbon molecules that immediately break down into three-carbon molecules called 3-phosphoglyceriacid, or 3-PGA. The ATP and NADPH made in the light reactions is used to convert the 3-PGA into glyceraldehyde-3-phosphate, or G3P sugar molecules. Most of the G3P molecules are recycled back into RuBP using energy from more ATP, but one out of every six produced leaves the cycle—the end product of the dark reactions.

Sugars and Starches

Glyceraldehyde-3-phosphate can double up to form larger sugar molecules like glucose and fructose. These molecules are processed, and from them, the still larger sucrose, a disaccharide commonly known as table sugar, is made, though this process takes place outside of the chloroplast, in the cytoplasm.

Alternatively, glucose monomers in the chloroplast can be linked together to make starch, which accumulates into the starch grains found in the chloroplast. Under conditions such as high atmospheriCO_2 concentrations, these starch grains may grow very large, distorting the grana and thylakoids. The starch granules displace the thylakoids, but leave them intact. Waterlogged roots can also cause starch buildup in the chloroplasts, possibly due to less sucrose being exported out of the chloroplast (or more accurately, the plant cell). This depletes a plant's free phosphate supply, which indirectly stimulates chloroplast starch synthesis. While linked to low photosynthesis rates, the starch grains themselves may not necessarily interfere significantly with the efficiency of photosynthesis, and might simply be a side effect of another photosynthesis-depressing factor.

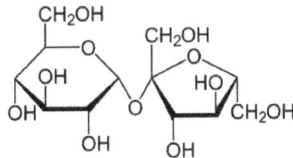

Sucrose is made up of a glucose monomer (left), and a fructose monomer (right).

Photorespiration

Photorespiration can occur when the oxygen concentration is too high. Rubisco cannot distinguish between oxygen and carbon dioxide very well, so it can accidentally add O_2 instead of CO_2 to RuBP. This process reduces the efficiency of photosynthesis—it consumes ATP and oxygen, releases CO_2, and produces no sugar. It can waste up to half the carbon fixed by the Calvin cycle. Several mechanisms have evolved in different lineages that raise the carbon dioxide concentration relative to oxygen within the chloroplast, increasing the efficiency of photosynthesis. These mechanisms are called carbon dioxide concentrating mechanisms, or CCMs. These include Crassulacean acid metabolism, C_4 carbon fixation, and pyrenoids. Chloroplasts in C_4 plants are notable as they exhibit a distinct chloroplast dimorphism.

pH

Because of the H^+ gradient across the thylakoid membrane, the interior of the thylakoid is acidic, with a pH around 4, while the stroma is slightly basic, with a pH of around 8. The optimal stroma pH for the Calvin cycle is 8.1, with the reaction nearly stopping when the pH falls below 7.3.

CO_2 in water can form carboniacid, which can disturb the pH of isolated chloroplasts, interfering with photosynthesis, even though CO_2 is used in photosynthesis. However, chloroplasts in living plant cells are not affected by this as much.

Chloroplasts can pump K^+ and H^+ ions in and out of themselves using a poorly understood light-driven transport system.

In the presence of light, the pH of the thylakoid lumen can drop up to 1.5 pH units, while the pH of the stroma can rise by nearly one pH unit.

Amino Acid Synthesis

Chloroplasts alone make almost all of a plant cell's amino acids in their stroma except the sulfur-containing ones like cysteine and methionine. Cysteine is made in the chloroplast (the proplastid too) but it is also synthesized in the cytosol and mitochondria, probably because it has trouble crossing membranes to get to where it is needed. The chloroplast is known to make the precursors to methionine but it is unclear whether the organelle carries out the last leg of the pathway or if it happens in the cytosol.

Other Nitrogen Compounds

Chloroplasts make all of a cell's purines and pyrimidines—the nitrogenous bases found in DNA and RNA. They also convert nitrite (NO_2^-) into ammonia (NH_3) which supplies the plant with nitrogen to make its amino acids and nucleotides.

Other Chemical Products

Chloroplasts are the site of complex lipid metabolism.

Differentiation, Replication, and Inheritance

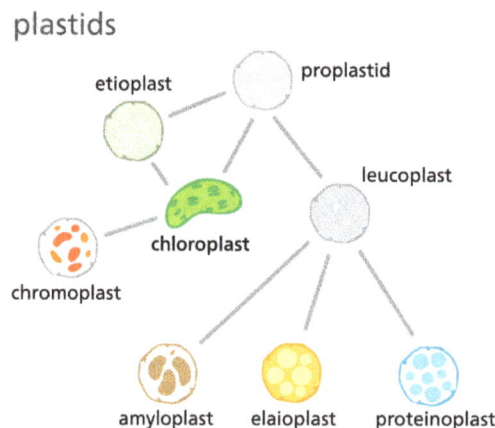

Plastid types (*Interactive diagram*) Plants contain many different kinds of plastids in their cells.

Chloroplasts are a special type of a plant cell organelle called a plastid, though the two terms are sometimes used interchangeably. There are many other types of plastids, which carry out various functions. All chloroplasts in a plant are descended from undifferentiated proplastids found in the zygote, or fertilized egg. Proplastids are commonly found in an adult plant's apical meristems. Chloroplasts do not normally develop from proplastids in root tip meristems—instead, the formation of starch-storing amyloplasts is more common.

In shoots, proplastids from shoot apical meristems can gradually develop into chloroplasts in photosynthetileaf tissues as the leaf matures, if exposed to the required light. This process involves invaginations of the inner plastid membrane, forming sheets of membrane that project into the internal stroma. These membrane sheets then fold to form thylakoids and grana.

If angiosperm shoots are not exposed to the required light for chloroplast formation, proplastids may develop into an etioplast stage before becoming chloroplasts. An etioplast is a plastid that lacks chlorophyll, and has inner membrane invaginations that form a lattice of tubes in their stroma, called a prolamellar body. While etioplasts lack chlorophyll, they have a yellow chlorophyll precursor stocked. Within a few minutes of light exposure, the prolamellar body begins to reorganize into stacks of thylakoids, and chlorophyll starts to be produced. This process, where the etioplast becomes a chloroplast, takes several hours. Gymnosperms do not require light to form chloroplasts.

Light, however, does not guarantee that a proplastid will develop into a chloroplast. Whether a proplastid develops into a chloroplast some other kind of plastid is mostly controlled by the nucleus and is largely influenced by the kind of cell it resides in.

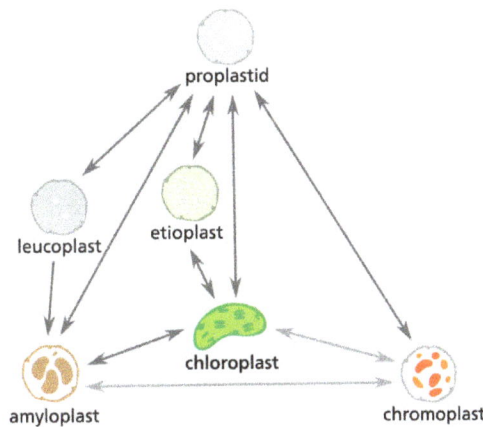

possible plastid interconversions

Many plastid interconversions are possible.

Plastid Interconversion

Plastid differentiation is not permanent, in fact many interconversions are possible. Chloroplasts may be converted to chromoplasts, which are pigment-filled plastids responsible for the bright colors seen in flowers and ripe fruit. Starch storing amyloplasts can also be converted to chromoplasts, and it is possible for proplastids to develop straight into chromoplasts. Chromoplasts and amyloplasts can also become chloroplasts, like what happens when a carrot or a potato is illuminated. If a plant is injured, or something else causes a plant cell to revert to a meristematistate, chloroplasts and other plastids can turn back into proplastids. Chloroplast, amyloplast, chromoplast, proplast, etc., are not absolute states—intermediate forms are common.

Chloroplast Division

Most chloroplasts in a photosyntheticell do not develop directly from proplastids or etioplasts. In fact, a typical shoot meristematiplant cell contains only 7–20 proplastids. These proplastids differentiate into chloroplasts, which divide to create the 30–70 chloroplasts found in a mature photosynthetiplant cell. If the cell divides, chloroplast division provides the additional chloroplasts to partition between the two daughter cells.

In single-celled algae, chloroplast division is the only way new chloroplasts are formed. There is no proplastid differentiation—when an algal cell divides, its chloroplast divides along with it, and each daughter cell receives a mature chloroplast.

Almost all chloroplasts in a cell divide, rather than a small group of rapidly dividing chloroplasts. Chloroplasts have no definite S-phase—their DNA replication is not synchronized or limited to that of their host cells. Much of what we know about chloroplast division comes from studying organisms like *Arabidopsis* and the red alga *Cyanidioschyzon merolæ*.

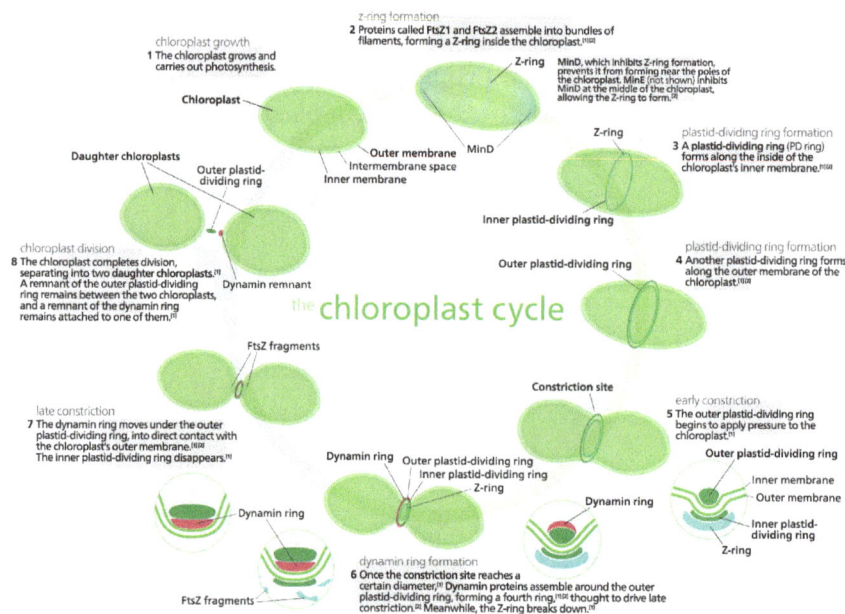

Most chloroplasts in plant cells, and all chloroplasts in algae arise from chloroplast division. *Picture references,*

The division process starts when the proteins FtsZ1 and FtsZ2 assemble into filaments, and with the help of a protein ARC6, form a structure called a Z-ring within the chloroplast's stroma. The Min system manages the placement of the Z-ring, ensuring that the chloroplast is cleaved more or less evenly. The protein MinD prevents FtsZ from linking up and forming filaments. Another protein ARC3 may also be involved, but it is not very well understood. These proteins are active at the poles of the chloroplast, preventing Z-ring formation there, but near the center of the chloroplast, MinE inhibits them, allowing the Z-ring to form.

Next, the two plastid-dividing rings, or PD rings form. The inner plastid-dividing ring is located in the inner side of the chloroplast's inner membrane, and is formed first. The outer plastid-dividing ring is found wrapped around the outer chloroplast membrane. It consists of filaments about 5 nanometers across, arranged in rows 6.4 nanometers apart, and shrinks to squeeze the chloroplast. This is when chloroplast constriction begins. In a few species like *Cyanidioschyzon merolæ*, chloroplasts have a third plastid-dividing ring located in the chloroplast's intermembrane space.

Late into the constriction phase, dynamin proteins assemble around the outer plastid-dividing ring, helping provide force to squeeze the chloroplast. Meanwhile, the Z-ring and the inner plastid-dividing ring break down. During this stage, the many chloroplast DNA plasmids floating around in the stroma are partitioned and distributed to the two forming daughter chloroplasts.

Later, the dynamins migrate under the outer plastid dividing ring, into direct contact with the chloroplast's outer membrane, to cleave the chloroplast in two daughter chloroplasts.

A remnant of the outer plastid dividing ring remains floating between the two daughter chloroplasts, and a remnant of the dynamin ring remains attached to one of the daughter chloroplasts.

Of the five or six rings involved in chloroplast division, only the outer plastid-dividing ring is present for the entire constriction and division phase—while the Z-ring forms first, constriction does not begin until the outer plastid-dividing ring forms.

Dividing chloroplasts

More dividing chloroplasts

Chloroplast division In this light micrograph of some moss chloroplasts, many dumbbell-shaped chloroplasts can be seen dividing. Grana are also just barely visible as small granules.

Regulation

In species of algae that contain a single chloroplast, regulation of chloroplast division is extremely important to ensure that each daughter cell receives a chloroplast—chloroplasts can't be made from scratch. In organisms like plants, whose cells contain multiple chloroplasts, coordination is looser and less important. It is likely that chloroplast and cell division are somewhat synchronized, though the mechanisms for it are mostly unknown.

Light has been shown to be a requirement for chloroplast division. Chloroplasts can grow and progress through some of the constriction stages under poor quality green light, but are slow to complete division—they require exposure to bright white light to complete division. Spinach leaves grown under green light have been observed to contain many large dumbbell-shaped chloroplasts. Exposure to white light can stimulate these chloroplasts to divide and reduce the population of dumbbell-shaped chloroplasts.

Chloroplast Inheritance

Like mitochondria, chloroplasts are usually inherited from a single parent. Biparental chloroplast inheritance—where plastid genes are inherited from both parent plants—occurs in very low levels in some flowering plants.

Many mechanisms prevent biparental chloroplast DNA inheritance, including selective destruction of chloroplasts or their genes within the gamete or zygote, and chloroplasts from one parent being excluded from the embryo. Parental chloroplasts can be sorted so that only one type is present in each offspring.

Gymnosperms, such as pine trees, mostly pass on chloroplasts paternally, while flowering plants often inherit chloroplasts maternally. Flowering plants were once thought to only inherit chloroplasts maternally. However, there are now many documented cases of angiosperms inheriting chloroplasts paternally.

Angiosperms, which pass on chloroplasts maternally, have many ways to prevent paternal inheritance. Most of them produce sperm cells that do not contain any plastids. There are many other documented mechanisms that prevent paternal inheritance in these flowering plants, such as different rates of chloroplast replication within the embryo.

Among angiosperms, paternal chloroplast inheritance is observed more often in hybrids than in offspring from parents of the same species. This suggests that incompatible hybrid genes might interfere with the mechanisms that prevent paternal inheritance.

Transplastomiplants

Recently, chloroplasts have caught attention by developers of genetically modified crops. Since, in most flowering plants, chloroplasts are not inherited from the male parent, transgenes in these plastids cannot be disseminated by pollen. This makes plastid transformation a valuable tool for the creation and cultivation of genetically modified plants that are biologically contained, thus posing significantly lower environmental risks. This biological containment strategy is therefore suitable for establishing the coexistence of conventional and organiagriculture. While the reliability of this mechanism has not yet been studied for all relevant crop species, recent results in tobacco plants are promising, showing a failed containment rate of transplastomiplants at 3 in 1,000,000.

Mutation

In biology, a mutation is the permanent alteration of the nucleotide sequence of the genome of an organism, virus, or extrachromosomal DNA or other genetielements. Mutations result from damage to DNA which then may undergo error-prone repair (especially microhomology-mediated end joining), or cause an error during other forms of repair, or else may cause an error during replication (translesion synthesis). Mutations may also result from insertion or deletion of segments of DNA due to mobile genetielements. Mutations may or may not produce discernible changes in the observable characteristics (phenotype) of an organism. Mutations play a part in both normal and abnormal biological processes including: evolution, cancer, and the development of the immune system, including junctional diversity.

Mutation can result in many different types of change in sequences. Mutations in genes can either have no effect, alter the product of a gene, or prevent the gene from functioning properly or completely. Mutations can also occur in nongeniregions. One study on genetivariations between different species of *Drosophila* suggests that, if a mutation changes a protein produced by a gene, the result is likely to be harmful, with an estimated 70 percent of amino acid polymorphisms that have damaging effects, and the remainder being either neutral or marginally beneficial. Due to the damaging effects that mutations can have on genes, organisms have mechanisms such as DNA repair to prevent or correct mutations by reverting the mutated sequence back to its original state.

Description

Mutations can involve the duplication of large sections of DNA, usually through genetirecombination. These duplications are a major source of raw material for evolving new genes, with tens to hundreds of genes duplicated in animal genomes every million years. Most genes belong to larger gene families of shared ancestry, known as homology. Novel genes are produced by several methods, commonly through the duplication and mutation of an ancestral gene, or by recombining parts of different genes to form new combinations with new functions.

Here, protein domains act as modules, each with a particular and independent function, that can be mixed together to produce genes encoding new proteins with novel properties. For example, the human eye uses four genes to make structures that sense light: three for cone cell or color vision and one for rod cell or night vision; all four arose from a single ancestral gene. Another advantage of duplicating a gene (or even an entire genome) is that this increases engineering redundancy; this allows one gene in the pair to acquire a new function while the other copy performs the original function. Other types of mutation occasionally create new genes from previously noncoding DNA.

Changes in chromosome number may involve even larger mutations, where segments of the DNA within chromosomes break and then rearrange. For example, in the Homininae, two chromosomes fused to produce human chromosome 2; this fusion did not occur in the lineage of the other apes, and they retain these separate chromosomes. In evolution, the most important role of such chromosomal rearrangements may be to accelerate the divergence of a population into new species by making populations less likely to interbreed, thereby preserving genetidifferences between these populations.

Sequences of DNA that can move about the genome, such as transposons, make up a major fraction of the genetimaterial of plants and animals, and may have been important in the evolution of genomes. For example, more than a million copies of the Alu sequence are present in the human genome, and these sequences have now been recruited to perform functions such as regulating gene expression. Another effect of these mobile DNA sequences is that when they move within a genome, they can mutate or delete existing genes and thereby produce genetidiversity.

Nonlethal mutations accumulate within the gene pool and increase the amount of genetivariation. The abundance of some genetichanges within the gene pool can be reduced by natural selection, while other "more favorable" mutations may accumulate and result in adaptive changes.

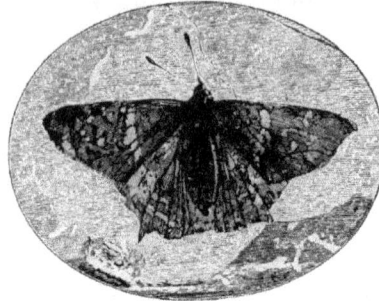

Prodryas persephone, a Late Eocene butterfly

For example, a butterfly may produce offspring with new mutations. The majority of these mutations will have no effect; but one might change the color of one of the butterfly's offspring, making it harder (or easier) for predators to see. If this color change is advantageous, the chance of this butterfly's surviving and producing its own offspring are a little better, and over time the number of butterflies with this mutation may form a larger percentage of the population.

Neutral mutations are defined as mutations whose effects do not influence the fitness of an individual. These can accumulate over time due to genetidrift. It is believed that the overwhelming majority of mutations have no significant effect on an organism's fitness. Also, DNA repair mechanisms are able to mend most changes before they become permanent mutations, and many organisms have mechanisms for eliminating otherwise-permanently mutated somaticells.

Beneficial mutations can improve reproductive success.

Causes

Four classes of mutations are (1) spontaneous mutations (molecular decay), (2) mutations due to error-prone replication bypass of naturally occurring DNA damage (also called error-prone translesion synthesis), (3) errors introduced during DNA repair, and (4) induced mutations caused by mutagens. Scientists may also deliberately introduce mutant sequences through DNA manipulation for the sake of scientifiexperimentation.

Spontaneous Mutation

Spontaneous mutations on the molecular level can be caused by:

- Tautomerism — A base is changed by the repositioning of a hydrogen atom, altering the hydrogen bonding pattern of that base, resulting in incorrect base pairing during replication.

- Depurination — Loss of a purine base (A or G) to form an apurinisite (AP site).

- Deamination — Hydrolysis changes a normal base to an atypical base containing a keto group in place of the original amine group. Examples include □ U and A □ HX (hypoxanthine), which can be corrected by DNA repair mechanisms; and 5Me(5-methylcytosine) □ T, which is less likely to be detected as a mutation because thymine is a normal DNA base.

- Slipped strand mispairing — Denaturation of the new strand from the template during replication, followed by renaturation in a different spot ("slipping"). This can lead to insertions or deletions.

Error-prone Replication Bypass

There is increasing evidence that the majority of spontaneously arising mutations are due to error-prone replication (translesion synthesis) past DNA damage in the template strand. Naturally occurring oxidative DNA damages arise at least 10,000 times per cell per day in humans and 50,000 times or more per cell per day in rats. In mice, the majority of mutations are caused by translesion synthesis. Likewise, in yeast, Kunz et al. found that more than 60% of the spontaneous single base pair substitutions and deletions were caused by translesion synthesis.

Errors Introduced During DNA Repair

Although naturally occurring double-strand breaks occur at a relatively low frequency in DNA, their repair often causes mutation. Non-homologous end joining (NHEJ) is a major pathway for repairing double-strand breaks. NHEJ involves removal of a few nucleotides to allow somewhat inaccurate alignment of the two ends for rejoining followed by addition of nucleotides to fill in gaps. As a consequence, NHEJ often introduces mutations.

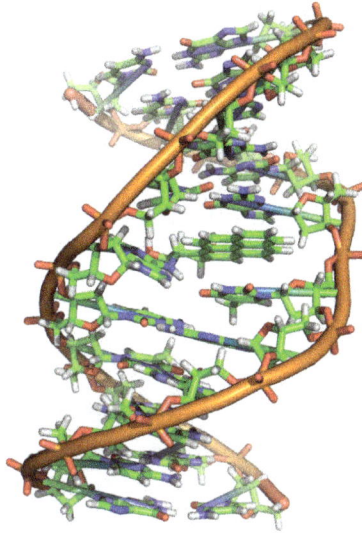

A covalent adduct between the metabolite of benzo[a]pyrene, the major mutagen in tobacco smoke, and DNA

Induced Mutation

Induced mutations on the molecular level can be caused by:-

- Chemicals

    - Hydroxylamine

    - Base analogs (e.g., Bromodeoxyuridine (BrdU))

    - Alkylating agents (e.g., *N*-ethyl-*N*-nitrosourea (ENU)). These agents can mutate both replicating and non-replicating DNA. In contrast, a base analog can mutate the DNA only when the analog is incorporated in replicating the DNA. Each of these classes of chemical mutagens has certain effects that then lead to transitions, transversions, or deletions.

- Agents that form DNA adducts (e.g., ochratoxin A)

- DNA intercalating agents (e.g., ethidium bromide)

- DNA crosslinkers

- Oxidative damage

- Nitrous acid converts amine groups on A and to diazo groups, altering their hydrogen bonding patterns, which leads to incorrect base pairing during replication.

- Radiation

  - Ultraviolet light (UV) (non-ionizing radiation). Two nucleotide bases in DNA—cytosine and thymine—are most vulnerable to radiation that can change their properties. UV light can induce adjacent pyrimidine bases in a DNA strand to become covalently joined as a pyrimidine dimer. UV radiation, in particular longer-wave UVA, can also cause oxidative damage to DNA.

Classification of Mutation Types

By Effect on Structure

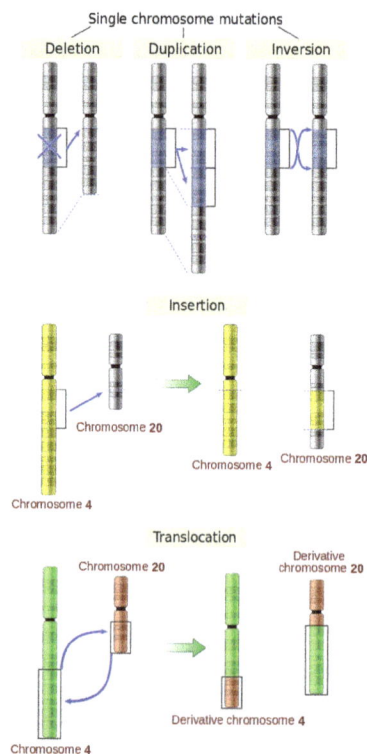

Illustrations of five types of chromosomal mutations.

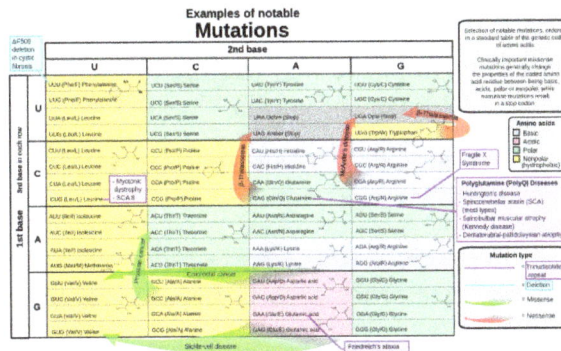

Selection of disease-causing mutations, in a standard table of the geneticode of amino acids.

The sequence of a gene can be altered in a number of ways. Gene mutations have varying effects on health depending on where they occur and whether they alter the function of essential proteins. Mutations in the structure of genes can be classified as:

- Small-scale mutations, such as those affecting a small gene in one or a few nucleotides, including:

  - Substitution mutations, often caused by chemicals or malfunction of DNA replication, exchange a single nucleotide for another. These changes are classified as transitions or transversions. Most common is the transition that exchanges a purine for a purine (A □ G) or a pyrimidine for a pyrimidine, (□ T). A transition can be caused by nitrous acid, base mis-pairing, or mutagenibase analogs such as BrdU. Less common is a transversion, which exchanges a purine for a pyrimidine or a pyrimidine for a purine (C/T □ A/G). An example of a transversion is the conversion of adenine (A) into a cytosine (C). A point mutation can be reversed by another point mutation, in which the nucleotide is changed back to its original state (true reversion) or by second-site reversion (a complementary mutation elsewhere that results in regained gene functionality). Point mutations that occur within the protein coding region of a gene may be classified into three kinds, depending upon what the erroneous codon codes for:

    - Silent mutations, which code for the same (or a sufficiently similar) amino acid.

    - Missense mutations, which code for a different amino acid.

    - Nonsense mutations, which code for a stop codon and can truncate the protein.

  - Insertions add one or more extra nucleotides into the DNA. They are usually caused by transposable elements, or errors during replication of repeating elements. Insertions in the coding region of a gene may alter splicing of the mRNA (splice site mutation), or cause a shift in the reading frame (frameshift), both of which can significantly alter the gene product. Insertions can be reversed by excision of the transposable element.

  - Deletions remove one or more nucleotides from the DNA. Like insertions, these mutations can alter the reading frame of the gene. In general, they are irreversible: Though exactly the same sequence might in theory be restored by an insertion, transposable elements able to revert a very short deletion (say 1–2 bases) in any location either are highly unlikely to exist or do not exist at all.

- Large-scale mutations in chromosomal structure, including:

  - Amplifications (or gene duplications) leading to multiple copies of all chromosomal regions, increasing the dosage of the genes located within them.

  - Deletions of large chromosomal regions, leading to loss of the genes within those regions.

  - Mutations whose effect is to juxtapose previously separate pieces of DNA, potentially bringing together separate genes to form functionally distinct fusion genes (e.g., bcr-abl). These include:

    - Chromosomal translocations: interchange of genetiparts from nonhomologous chromosomes.

    - Interstitial deletions: an intra-chromosomal deletion that removes a segment of DNA from a single chromosome, thereby apposing previously distant genes. For example, cells isolated from a human astrocytoma, a type of brain tumor, were found to have a chromosomal deletion removing sequences between the Fused in Glioblastoma (FIG) gene and the receptor tyrosine kinase (ROS), producing a fusion protein (FIG-ROS). The abnormal FIG-ROS fusion protein has constitutively active kinase activity that causes oncogenitransformation (a transformation from normal cells to cancer cells).

    - Chromosomal inversions: reversing the orientation of a chromosomal segment.

  - Loss of heterozygosity: loss of one allele, either by a deletion or a genetirecombination event, in an organism that previously had two different alleles.

By Effect on Function

- Loss-of-function mutations, also called inactivating mutations, result in the gene product having less or no function (being partially or wholly inactivated). When the allele has a complete loss of function (null allele), it is often called an amorphimutation in the Muller's morphs schema. Phenotypes associated with such mutations are most often recessive. Exceptions are when the organism is haploid, or when the reduced dosage of a normal gene product is not enough for a normal phenotype (this is called haploinsufficiency).

- Gain-of-function mutations, also called activating mutations, change the gene product such that its effect gets stronger (enhanced activation) or even is superseded by a different and abnormal function. When the new allele is created, a heterozygote containing the newly created allele as well as the original will express the new allele; genetically this defines the mutations as dominant phenotypes. Often called a neomorphimutation.

- Dominant negative mutations (also called antimorphimutations) have an altered gene product that acts antagonistically to the wild-type allele. These mutations usually result in an altered molecular function (often inactive) and are characterized by a dominant or semi-dominant phenotype. In humans, dominant negative mutations have been implicated in cancer (e.g., mutations in genes p53, ATM, CEBPA and PPARgamma). Marfan syndrome is caused by mutations in the *FBN1* gene, located on chromosome 15, which encodes fibrillin-1, a glycoprotein component of the extracellular matrix. Marfan syndrome is also an example of dominant negative mutation and haploinsufficiency.

- Lethal mutations are mutations that lead to the death of the organisms that carry the mutations.

- A back mutation or reversion is a point mutation that restores the original sequence and hence the original phenotype.

By Effect on Fitness

In applied genetics, it is usual to speak of mutations as either harmful or beneficial.

- A harmful, or deleterious, mutation decreases the fitness of the organism.

- A beneficial, or advantageous mutation increases the fitness of the organism. Mutations that promotes traits that are desirable, are also called beneficial. In theoretical population genetics, it is more usual to speak of mutations as deleterious or advantageous than harmful or beneficial.

- A neutral mutation has no harmful or beneficial effect on the organism. Such mutations occur at a steady rate, forming the basis for the molecular clock. In the neutral theory of molecular evolution, neutral mutations provide genetidrift as the basis for most variation at the molecular level.

- A nearly neutral mutation is a mutation that may be slightly deleterious or advantageous, although most nearly neutral mutations are slightly deleterious.

Distribution of Fitness Effects

Attempts have been made to infer the distribution of fitness effects (DFE) using mutagenesis experiments and theoretical models applied to molecular sequence data. DFE, as used to determine the relative abundance of different types of mutations (i.e., strongly deleterious, nearly neutral or advantageous), is relevant to many evolutionary questions, such as the maintenance of genetivariation, the rate of genomidecay, the maintenance of outcrossing sexual reproduction as opposed to inbreeding and the evolution of sex and genetirecombination. In summary, the DFE plays an important role in predicting evolutionary dynamics. A variety of approaches have been used to study the DFE, including theoretical, experimental and analytical methods.

- Mutagenesis experiment: The direct method to investigate the DFE is to induce mutations and then measure the mutational fitness effects, which has already been done in viruses, bacteria, yeast, and *Drosophila*. For example, most studies of the DFE in viruses used site-directed mutagenesis to create point mutations and measure relative fitness of each mutant. In *Escherichia coli*, one study used transposon mutagenesis to directly measure the fitness of a random insertion of a derivative of Tn10. In yeast, a combined mutagenesis and deep sequencing approach has been developed to generate high-quality systematimutant libraries and measure fitness in high throughput. However, given that many mutations have effects too small to be detected and that mutagenesis experiments can detect only mutations of moderately large effect; DNA sequence data analysis can provide valuable information about these mutations.

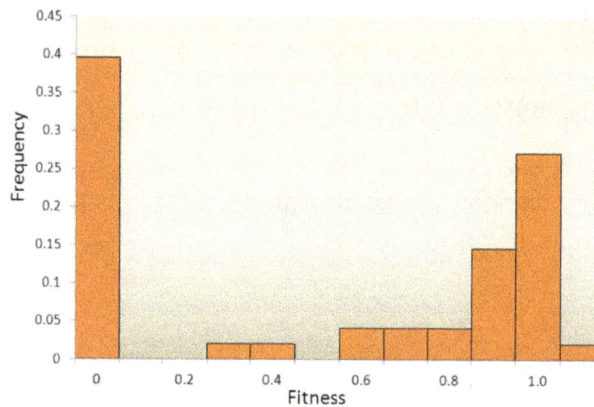

The distribution of fitness effects (DFE) of mutations in vesicular stomatitis virus. In this experiment, random mutations were introduced into the virus by site-directed mutagenesis, and the fitness of each mutant was compared with the ancestral type. A fitness of zero, less than one, one, more than one, respectively, indicates that mutations are lethal, deleterious, neutral, and advantageous.

- Molecular sequence analysis: With rapid development of DNA sequencing technology, an enormous amount of DNA sequence data is available and even more is forthcoming in the future. Various methods have been developed to infer the DFE from DNA sequence data. By examining DNA sequence differences within and between species, we are able to infer various characteristics of the DFE for neutral, deleterious and advantageous mutations. To be specific, the DNA sequence analysis approach allows us to estimate the effects of mutations with very small effects, which are hardly detectable through mutagenesis experiments.

One of the earliest theoretical studies of the distribution of fitness effects was done by Motoo Kimura, an influential theoretical population geneticist. His neutral theory of molecular evolution proposes that most novel mutations will be highly deleterious, with a small fraction being neutral. Hiroshi Akashi more recently proposed a bimodal model for the DFE, with modes centered around highly deleterious and neutral mutations. Both theories agree that the vast majority of novel mutations are neutral or deleterious and that advantageous mutations are rare, which has been supported by experimental results. One example is a study done on the DFE of random mutations in vesicular stomatitis virus. Out of all mutations, 39.6% were lethal, 31.2% were non-lethal deleterious, and 27.1% were neutral. Another example comes from a high throughput mutagenesis experiment with yeast. In this experiment it was shown that the overall DFE is bimodal, with a cluster of neutral mutations, and a broad distribution of deleterious mutations.

Though relatively few mutations are advantageous, those that are play an important role in evolutionary changes. Like neutral mutations, weakly selected advantageous mutations can be lost due to random genetidrift, but strongly selected advantageous mutations are more likely to be fixed. Knowing the DFE of advantageous mutations may lead to increased ability to predict the evolutionary dynamics. Theoretical work on the DFE for advantageous mutations has been done by John H. Gillespie and H. Allen Orr. They proposed that the distribution for advantageous mutations should be exponential under a wide range of conditions, which, in general, has been supported by experimental studies, at least for strongly selected advantageous mutations.

In general, it is accepted that the majority of mutations are neutral or deleterious, with rare mutations being advantageous; however, the proportion of types of mutations varies between species.

This indicates two important points: first, the proportion of effectively neutral mutations is likely to vary between species, resulting from dependence on effective population size; second, the average effect of deleterious mutations varies dramatically between species. In addition, the DFE also differs between coding regions and noncoding regions, with the DFE of noncoding DNA containing more weakly selected mutations.

By Impact on Protein Sequence

- A frameshift mutation is a mutation caused by insertion or deletion of a number of nucleotides that is not evenly divisible by three from a DNA sequence. Due to the triplet nature of gene expression by codons, the insertion or deletion can disrupt the reading frame, or the grouping of the codons, resulting in a completely different translation from the original. The earlier in the sequence the deletion or insertion occurs, the more altered the protein produced is.

- In contrast, any insertion or deletion that is evenly divisible by three is termed an *in-frame mutation*

- A nonsense mutation is a point mutation in a sequence of DNA that results in a premature stop codon, or a *nonsense codon* in the transcribed mRNA, and possibly a truncated, and often nonfunctional protein product.

- Missense mutations or *nonsynonymous mutations* are types of point mutations where a single nucleotide is changed to cause substitution of a different amino acid. This in turn can render the resulting protein nonfunctional. Such mutations are responsible for diseases such as Epidermolysis bullosa, sickle-cell disease, and SOD1-mediated ALS.

- A neutral mutation is a mutation that occurs in an amino acid codon that results in the use of a different, but chemically similar, amino acid. The similarity between the two is enough that little or no change is often rendered in the protein. For example, a change from AAA to AGA will encode arginine, a chemically similar molecule to the intended lysine.

- Silent mutations are mutations that do not result in a change to the amino acid sequence of a protein, unless the changed amino acid is sufficiently similar to the original. They may occur in a region that does not code for a protein, or they may occur within a codon in a manner that does not alter the final amino acid sequence. The phrase *silent mutation* is often used interchangeably with the phrase *synonymous mutation*; however, synonymous mutations are a subcategory of the former, occurring only within exons (and necessarily exactly preserving the amino acid sequence of the protein). Synonymous mutations occur due to the degenerate nature of the geneticode.

By Inheritance

In multicellular organisms with dedicated reproductive cells, mutations can be subdivided into germline mutations, which can be passed on to descendants through their reproductive cells, and somatimutations (also called acquired mutations), which involve cells outside the dedicated reproductive group and which are not usually transmitted to descendants.

A mutation has caused this garden moss rose to produce flowers of different colors. This is a somatimutation that may also be passed on in the germline.

A germline mutation gives rise to a *constitutional mutation* in the offspring, that is, a mutation that is present in every cell. A constitutional mutation can also occur very soon after fertilisation, or continue from a previous constitutional mutation in a parent.

The distinction between germline and somatimutations is important in animals that have a dedicated germline to produce reproductive cells. However, it is of little value in understanding the effects of mutations in plants, which lack dedicated germline. The distinction is also blurred in those animals that reproduce asexually through mechanisms such as budding, because the cells that give rise to the daughter organisms also give rise to that organism's germline. A new germline mutation that was not inherited from either parent is called a *de novo* mutation.

Diploid organisms (e.g., humans) contain two copies of each gene—a paternal and a maternal allele. Based on the occurrence of mutation on each chromosome, we may classify mutations into three types.

- A heterozygous mutation is a mutation of only one allele.

- A homozygous mutation is an identical mutation of both the paternal and maternal alleles.

- Compound heterozygous mutations or a geneticompound comprises two different mutations in the paternal and maternal alleles.

A wild type or homozygous non-mutated organism is one in which neither allele is mutated.

Special Classes

- Conditional mutation is a mutation that has wild-type (or less severe) phenotype under certain "permissive" environmental conditions and a mutant phenotype under certain "restrictive" conditions. For example, a temperature-sensitive mutation can cause cell death at high temperature (restrictive condition), but might have no deleterious consequences at a lower temperature (permissive condition).

- Replication Timing Quantitative Trait Loci Affects DNA Replication.

Nomenclature

In order to categorize a mutation as such, the "normal" sequence must be obtained from the DNA of a "normal" or "healthy" organism (as opposed to a "mutant" or "sick" one), it should be identified and reported; ideally, it should be made publicly available for a straightforward nucleotide-by-nucleotide comparison, and agreed upon by the scientificommunity or by a group of expert geneticists and biologists, who have the responsibility of establishing the *standard* or so-called "consensus" sequence. This step requires a tremendous scientifieffort. Once the consensus sequence is known, the mutations in a genome can be pinpointed, described, and classified. The committee of the Human Genome Variation Society (HGVS) has developed the standard human sequence variant nomenclature, which should be used by researchers and DNA diagnosticenters to generate unambiguous mutation descriptions. In principle, this nomenclature can also be used to describe mutations in other organisms. The nomenclature specifies the type of mutation and base or amino acid changes.

- Nucleotide substitution (e.g., 76A>T) — The number is the position of the nucleotide from the 5' end; the first letter represents the wild-type nucleotide, and the second letter represents the nucleotide that replaced the wild type. In the given example, the adenine at the 76th position was replaced by a thymine.

  - If it becomes necessary to differentiate between mutations in genomiDNA, mitochondrial DNA, and RNA, a simple convention is used. For example, if the 100th base of a nucleotide sequence mutated from G to C, then it would be written as g.100G>if the mutation occurred in genomiDNA, m.100G>if the mutation occurred in mitochondrial DNA, or r.100g>if the mutation occurred in RNA. Note that, for mutations in RNA, the nucleotide code is written in lower case.

- Amino acid substitution (e.g., D111E) — The first letter is the one letter code of the wild-type amino acid, the number is the position of the amino acid from the N-terminus, and the second letter is the one letter code of the amino acid present in the mutation. Nonsense mutations are represented with an X for the second amino acid (e.g. D111X).

- Amino acid deletion (e.g., ΔF508) — The Greek letter Δ (delta) indicates a deletion. The letter refers to the amino acid present in the wild type and the number is the position from the N terminus of the amino acid were it to be present as in the wild type.

Mutation Rates

Mutation rates vary substantially across species, and the evolutionary forces that generally determine mutation is the subject of ongoing investigation.

Harmful Mutations

Changes in DNA caused by mutation can cause errors in protein sequence, creating partially or completely non-functional proteins. Each cell, in order to function correctly, depends on thousands of proteins to function in the right places at the right times. When a mutation alters a protein that plays a critical role in the body, a medical condition can result. A condition caused by mutations in one or more genes is called a genetidisorder. Some mutations alter a gene's DNA base sequence but do not change the function of the protein made by the gene. One study on the comparison of genes between different species of *Drosophila* suggests that if a mutation does change a protein, this will probably be harmful, with an estimated 70 percent of amino acid polymorphisms having damaging effects, and the remainder being either neutral or weakly beneficial. Studies have shown that only 7% of point mutations in noncoding DNA of yeast are deleterious and 12% in coding DNA are deleterious. The rest of the mutations are either neutral or slightly beneficial.

If a mutation is present in a germ cell, it can give rise to offspring that carries the mutation in all of its cells. This is the case in hereditary diseases. In particular, if there is a mutation in a DNA repair gene within a germ cell, humans carrying such germline mutations may have an increased risk of cancer. A list of 34 such germline mutations is given in the article DNA repair-deficiency disorder. An example of one is albinism, a mutation that occurs in the OCA1 or OCA2 gene. Individuals with this disorder are more prone to many types of cancers, other disorders and have impaired vision. On the other hand, a mutation may occur in a somaticell of an organism. Such mutations will be present in all descendants of this cell within the same organism, and certain mutations can cause the cell to become malignant, and, thus, cause cancer.

A DNA damage can cause an error when the DNA is replicated, and this error of replication can cause a gene mutation that, in turn, could cause a genetidisorder. DNA damages are repaired by the DNA repair system of the cell. Each cell has a number of pathways through which enzymes recognize and repair damages in DNA. Because DNA can be damaged in many ways, the process of DNA repair is an important way in which the body protects itself from disease. Once DNA damage has given rise to a mutation, the mutation cannot be repaired. DNA repair pathways can only recognize and act on "abnormal" structures in the DNA. Once a mutation occurs in a gene sequence it then has normal DNA structure and cannot be repaired.

Beneficial Mutations

Although mutations that cause changes in protein sequences can be harmful to an organism, on occasions the effect may be positive in a given environment. In this case, the mutation may enable the mutant organism to withstand particular environmental stresses better than wild-type organisms, or reproduce more quickly. In these cases a mutation will tend to become more common in a population through natural selection.

For example, a specifi32 base pair deletion in human CCR5 (CCR5-Δ32) confers HIV resistance to homozygotes and delays AIDS onset in heterozygotes. One possible explanation of the etiology of the relatively high frequency of CCR5-Δ32 in the European population is that it conferred resistance to the buboniplague in mid-14th century Europe. People with this mutation were more likely to survive infection; thus its frequency in the population increased. This theory could explain why this mutation is not found in Southern Africa, which remained untouched by buboniplague. A newer theory suggests that the selective pressure on the CCR5 Delta 32 mutation was caused by smallpox instead of the buboniplague.

Another example is sickle-cell disease, a blood disorder in which the body produces an abnormal type of the oxygen-carrying substance hemoglobin in the red blood cells. One-third of all indigenous inhabitants of Sub-Saharan Africa carry the gene, because, in areas where malaria is common, there is a survival value in carrying only a single sickle-cell gene (sickle cell trait).

Those with only one of the two alleles of the sickle-cell disease are more resistant to malaria, since the infestation of the malaria *Plasmodium* is halted by the sickling of the cells that it infests.

Prion Mutations

Prions are proteins and do not contain genetimaterial. However, prion replication has been shown to be subject to mutation and natural selection just like other forms of replication.

Somatimutations

A change in the genetistructure that is not inherited from a parent, and also not passed to off-spring, is called a *somaticell genetimutation* or *acquired mutation*.

Cells with heterozygous mutations (one good copy of gene and one mutated copy) may function normally with the unmutated copy until the good copy has been spontaneously somatically mutated. This kind of mutation happens all the time in living organisms, but it is difficult to measure the rate. Measuring this rate is important in predicting the rate at which people may develop cancer.

Point mutations may arise from spontaneous mutations that occur during DNA replication. The rate of mutation may be increased by mutagens. Mutagens can be physical, such as radiation from UV rays, X-rays or extreme heat, or chemical (molecules that misplace base pairs or disrupt the helical shape of DNA). Mutagens associated with cancers are often studied to learn about cancer and its prevention.

Germline

Cormlets of *Watsonia meriana*, an example of apomixis

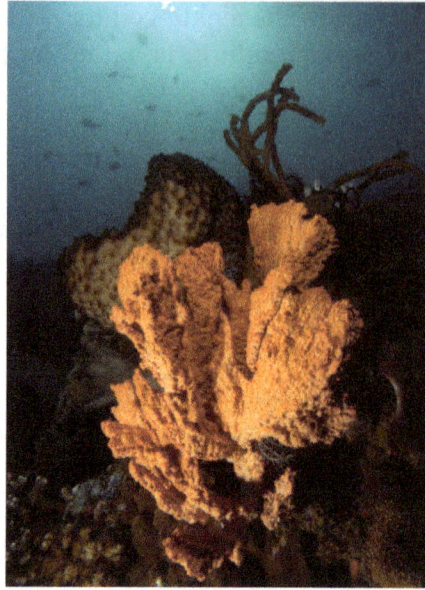

Clathria tuberosa, an example of a sponge that can grow indefinitely from somatitissue and reconstitute itself from totipotent separated somaticells

In biology and genetics, the germline in a multicellular organism is that population of its bodily cells that are so differentiated or segregated that in the usual processes of reproduction they may pass on their genetimaterial to the progeny.

As a rule this passing-on happens via a process of *sexual reproduction*; typically it is a process that includes systematichanges to the genetimaterial, changes that arise during recombination, meiosis and fertilization or syngamy for example. However, there are many exceptions, including processes and concepts such as various forms of apomixis, autogamy, automixis, cloning, or parthenogenesis. The cells of the germline commonly are called germ cells.

For example, gametes such as the sperm or the egg are part of the germline. So are the cells that divide to produce the gametes, called gametocytes, the cells that produce those, called gametogonia, and all the way back to the zygote, the cell from which the individual developed.

In sexually reproducing organisms, cells that are not in the germline are called somaticells. The term refers to all of the cells of body apart from the gametes. According to this view mutations, recombinations and other genetichanges in the germline may be passed to offspring, but a change in a somaticell will not be. This need not apply to somatically reproducing organisms, such as some Porifera and many plants. For example, many varieties of citrus, plants in the Rosaceae and some in the Asteraceae, such as *Taraxacum* produce seeds apomictically when somatidiploid cells displace the ovule or early embryo.

As August Weismann proposed and pointed out, a germline cell is immortal in the sense that it is part of a lineage that has reproduced indefinitely since the beginning of life and, barring accident could continue doing so indefinitely. Somaticells of most organisms however, can only approach any such capability to a limited extent and under special conditions. It is now known in some detail that this distinction between somatiand germ cells is partly artificial and depends on particular circumstances and internal cellular mechanisms such as telomeres and controls such as the selective application of telomerase in germ cells, stem cells and the like. Weismann however worked long before such mechanisms were known, let alone epigenetimechanisms or even the genetirole of chromosomes, and he believed that there was some clear qualitative difference between germ cells and somaticells, though he did realise that the somaticells differentiated from the germ cells. Many of his views necessarily changed during his life, and some of the resulting inconsistencies were discussed in depth by George Romanes. However Weismann was under no illusions concerning the limitations of his ideas in the absence of hard data concerning the nature of the systems he was speculating on or studying, and he discussed the limitations frankly and analytically.

Not all multicellular organisms differentiate into somatiand germ lines, but in the absence of specialised technical human intervention practically all but the simplest multicellular structures do so. In such organisms somaticells tend to be practically totipotent, and for over a century sponge cells have been known to reassemble into new sponges after having been separated by forcing them through a sieve.

Germline can refer to a lineage of cells spanning many generations of individuals—for example, the germline that links any living individual to the hypothetical last universal ancestor, from which all plants and animals descend.

Polyploid

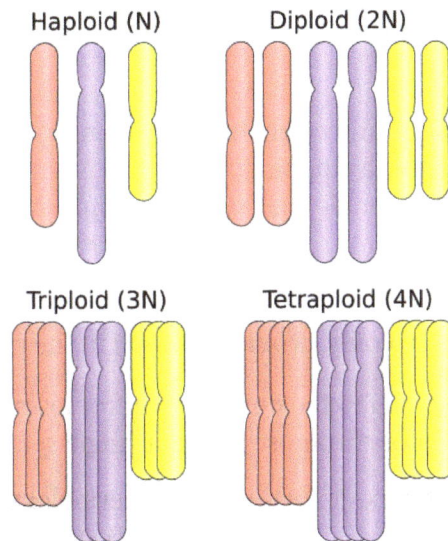

This image shows haploid (single), diploid (double), triploid (triple), and tetraploid (quadruple) sets of chromosomes. Triploid and tetraploid chromosomes are examples of polyploidy.

Polyploid cells and organisms are those containing more than two paired (homologous) sets of chromosomes. Most species whose cells have nuclei (Eukaryotes) are diploid, meaning they have two sets of chromosomes—one set inherited from each parent. However, polyploidy is found in some organisms and is especially common in plants. In addition, polyploidy occurs in some tissues of animals that are otherwise diploid, such as human muscle tissues. This is known as endopolyploidy. Species whose cells do not have nuclei, that is, Prokaryotes, may be polyploid organisms, as seen in the large bacterium *Epulopiscium fishelsoni* . Hence ploidy is defined with respect to a cell. Most eukaryotes have diploid somaticells, but produce haploid gametes (eggs and sperm) by meiosis. A monoploid has only one set of chromosomes, and the term is usually only applied to cells or organisms that are normally diploid. Male bees and other Hymenoptera, for example, are monoploid. Unlike animals, plants and multicellular algae have life cycles with two alternating multicellular generations. The gametophyte generation is haploid, and produces gametes by mitosis, the sporophyte generation is diploid and produces spores by meiosis.

Polyploidy refers to a numerical change in a whole set of chromosomes. Organisms in which a particular chromosome, or chromosome segment, is under- or overrepresented are said to be aneuploid (from the Greek words meaning "not", "good", and "fold"). Therefore, the distinction between aneuploidy and polyploidy is that aneuploidy refers to a numerical change in part of the chromosome set, whereas polyploidy refers to a numerical change in the whole set of chromosomes.

Polyploidy may occur due to abnormal cell division, either during mitosis, or commonly during metaphase I in meiosis.

Polyploidy occurs in highly differentiated human tissues in the liver, heart muscle and bone marrow. It occurs in the somaticells of some animals, such as goldfish, salmon, and salamanders, but is especially common among ferns and flowering plants, including both wild and cultivated species. Wheat, for example, after millennia of hybridization and modification by humans, has strains that are diploid (two sets of chromosomes), tetraploid (four sets of chromosomes) with the common name of durum or macaroni wheat, and hexaploid (six sets of chromosomes) with the common name of bread wheat. Many agriculturally important plants of the genus *Brassica* are also tetraploids.

Polyploidy can be induced in plants and cell cultures by some chemicals: the best known is colchicine, which can result in chromosome doubling, though its use may have other less obvious consequences as well. Oryzalin will also double the existing chromosome content.

Types

Organ-specifipatterns of endopolyploidy (from $2x$ to $64x$) in the giant ant *Dinoponera australis*

Polyploid types are labeled according to the number of chromosome sets in the nucleus. The letter x is used to represent the number of chromosomes in a single set.

- triploid (three sets; $3x$), for example seedless watermelons, common in the phylum Tardigrada
- tetraploid (four sets; $4x$), for example Salmonidae fish, the cotton *Gossypium hirsutum*
- pentaploid (five sets; $5x$), for example Kenai Birch (*Betula papyrifera* var. *kenaica*)
- hexaploid (six sets; $6x$), for example wheat, kiwifruit
- heptaploid or septaploid (seven sets; $7x$)
- octaploid or octoploid, (eight sets; $8x$), for example *Acipenser* (genus of sturgeon fish), dahlias

- decaploid (ten sets; 10x), for example certain strawberries

- dodecaploid (twelve sets; 12x), for example the plant *Celosia argentea* or the invasive one *Spartina anglica* or the amphibian *Xenopus ruwenzoriensis*.

Animals

Examples in animals are more common in non-vertebrates such as flatworms, leeches, and brine shrimp. Within vertebrates, examples of stable polyploidy include the salmonids and many cyprinids (i.e. carp). Some fish have as many as 400 chromosomes. Polyploidy also occurs commonly in amphibians; for example the biomedically-important *Xenopus* genus contains many different species with as many as 12 sets of chromosomes (dodecaploid). Polyploid lizards are also quite common, but are sterile and must reproduce by parthenogenesis. Polyploid mole salamanders (mostly triploids) are all female and reproduce by kleptogenesis, "stealing" spermatophores from diploid males of related species to trigger egg development but not incorporating the males' DNA into the offspring. While mammalian liver cells are polyploid, rare instances of polyploid mammals are known, but most often result in prenatal death.

An octodontid rodent of Argentina's harsh desert regions, known as the Plains Viscacha-Rat (*Tympanoctomys barrerae*) has been reported as an exception to this 'rule'. However, careful analysis using chromosome paints shows that there are only two copies of each chromosome in *T. barrerae*, not the four expected if it were truly a tetraploid. The rodent is not a rat, but kin to guinea pigs and chinchillas. Its "new" diploid [2n] number is 102 and so its cells are roughly twice normal size. Its closest living relation is *Octomys mimax*, the Andean Viscacha-Rat of the same family, whose 2n = 56. It was therefore surmised that an *Octomys*-like ancestor produced tetraploid (i.e., 2n = 4x = 112) offspring that were, by virtue of their doubled chromosomes, reproductively isolated from their parents.

Polyploidy was induced in fish by Har Swarup (1956) using a cold-shock treatment of the eggs close to the time of fertilization, which produced triploid embryos that successfully matured. Cold or heat shock has also been shown to result in unreduced amphibian gametes, though this occurs more commonly in eggs than in sperm. John Gurdon (1958) transplanted intact nuclei from somaticells to produce diploid eggs in the frog, *Xenopus* (an extension of the work of Briggs and King in 1952) that were able to develop to the tadpole stage. The British Scientist, J. B. S. Haldane hailed the work for its potential medical applications and, in describing the results, became one of the first to use the word "clone" in reference to animals. Later work by Shinya Yamanaka showed how mature cells can be reprogrammed to become pluripotent, extending the possibilities to non-stem cells. Gurdon and Yamanaka were jointly awarded the Nobel Prize in 2012 for this work.

Humans

True polyploidy rarely occurs in humans, although polyploid cells occur in highly differentiated tissue, such as liver parenchyma and heart muscle, and in bone marrow. Aneuploidy is more common.

Polyploidy occurs in humans in the form of triploidy, with 69 chromosomes (sometimes called 69,XXX), and tetraploidy with 92 chromosomes (sometimes called 92,XXXX). Triploidy, usually due to polyspermy, occurs in about 2–3% of all human pregnancies and ~15% of miscarriages. The vast majority of triploid conceptions end as a miscarriage; those that do survive to term typically die shortly after birth. In some cases, survival past birth may extend longer if there is mixoploidy with both a diploid and a triploid cell population present.

Triploidy may be the result of either digyny (the extra haploid set is from the mother) or diandry (the extra haploid set is from the father). Diandry is mostly caused by reduplication of the paternal haploid set from a single sperm, but may also be the consequence of dispermi(two sperm) fertilization of the egg. Digyny is most commonly caused by either failure of one meiotidivision during oogenesis leading to a diploid oocyte or failure to extrude one polar body from the oocyte. Diandry appears to predominate among early miscarriages, while digyny predominates among triploid zygotes that survive into the fetal period. However, among early miscarriages, digyny is also more common in those cases <8.5 weeks gestational age or those in which an embryo is present. There are also two distinct phenotypes in triploid placentas and fetuses that are dependent on the origin of the extra haploid set. In digyny, there is typically an asymmetripoorly grown fetus, with marked adrenal hypoplasia and a very small placenta. In diandry, a partial hydatidiform mole develops. These parent-of-origin effects reflect the effects of genomiimprinting.

Complete tetraploidy is more rarely diagnosed than triploidy, but is observed in 1–2% of early miscarriages. However, some tetraploid cells are commonly found in chromosome analysis at prenatal diagnosis and these are generally considered 'harmless'. It is not clear whether these tetraploid cells simply tend to arise during *in vitro* cell culture or whether they are also present in placental cells *in vivo*. There are, at any rate, very few clinical reports of fetuses/infants diagnosed with tetraploidy mosaicism.

Mixoploidy is quite commonly observed in human preimplantation embryos and includes haploid/diploid as well as diploid/tetraploid mixed cell populations. It is unknown whether these embryos fail to implant and are therefore rarely detected in ongoing pregnancies or if there is simply a selective process favoring the diploid cells.

Plants

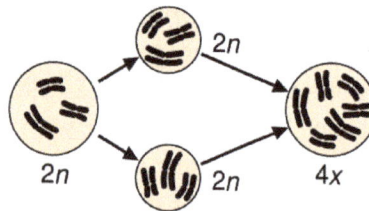

Speciation via polyploidy: A diploid cell undergoes failed meiosis, producing diploid gametes, which self-fertilize to produce a tetraploid zygote.

Polyploidy is pervasive in plants and some estimates suggest that 30–80% of living plant species are polyploid, and many lineages show evidence of ancient polyploidy (paleopolyploidy) in their genomes. Huge explosions in angiosperm species diversity appear to have coincided with the timing of ancient genome duplications shared by many species. It has been established that 15% of angiosperm and 31% of fern speciation events are accompanied by ploidy increase.

Polyploid plants can arise spontaneously in nature by several mechanisms, including meiotior mitotifailures, and fusion of unreduced (2n) gametes. Both autopolyploids (e.g. potato) and allopolyploids (e.g. canola, wheat, cotton) can be found among both wild and domesticated plant species.

Most polyploids display novel variation or morphologies relative to their parental species, that may contribute to the processes of speciation and eco-niche exploitation. The mechanisms leading to novel variation in newly formed allopolyploids may include gene dosage effects (resulting from more numerous copies of genome content), the reunion of divergent gene regulatory hierarchies, chromosomal rearrangements, and epigenetiremodeling, all of which affect gene content and/or expression levels. Many of these rapid changes may contribute to reproductive isolation and speciation. However seed generated from interploidy crosses, such as between polyploids and their parent species, usually suffer from aberrant endosperm development which impairs their viability, thus contributing to polyploid speciation.

Lomatia tasmanica is an extremely rare Tasmanian shrub that is triploid and sterile; reproduction is entirely vegetative, with all plants having the same geneticconstitution.

There are few naturally occurring polyploid conifers. One example is the Coast Redwood *Sequoia sempervirens*, which is a hexaploid (6x) with 66 chromosomes (2n = 6x = 66), although the origin is unclear.

Aquatiplants, especially the Monocotyledons, include a large number of polyploids.

Crops

The induction of polyploidy is a common technique to overcome the sterility of a hybrid species during plant breeding. For example, Triticale is the hybrid of wheat (*Triticum turgidum*) and rye (*Secale cereale*). It combines sought-after characteristics of the parents, but the initial hybrids are sterile. After polyploidization, the hybrid becomes fertile and can thus be further propagated to become triticale.

In some situations, polyploid crops are preferred because they are sterile. For example, many seedless fruit varieties are seedless as a result of polyploidy. Such crops are propagated using asexual techniques, such as grafting.

Polyploidy in crop plants is most commonly induced by treating seeds with the chemical colchicine.

Examples

- Triploid crops: apple, banana, citrus, ginger, watermelon
- Tetraploid crops: apple, durum or macaroni wheat, cotton, potato, canola/rapeseed, leek, tobacco, peanut, kinnow, Pelargonium
- Hexaploid crops: chrysanthemum, bread wheat, triticale, oat, kiwifruit
- Octaploid crops: strawberry, dahlia, pansies, sugar cane, oca (*Oxalis tuberosa*)
- Dodecaploid crops: some sugar cane hybrids

Some crops are found in a variety of ploidies: tulips and lilies are commonly found as both diploid and triploid; daylilies (*Hemerocallis* cultivars) are available as either diploid or tetraploid; apples and kinnows can be diploid, triploid, or tetraploid.

Fungi

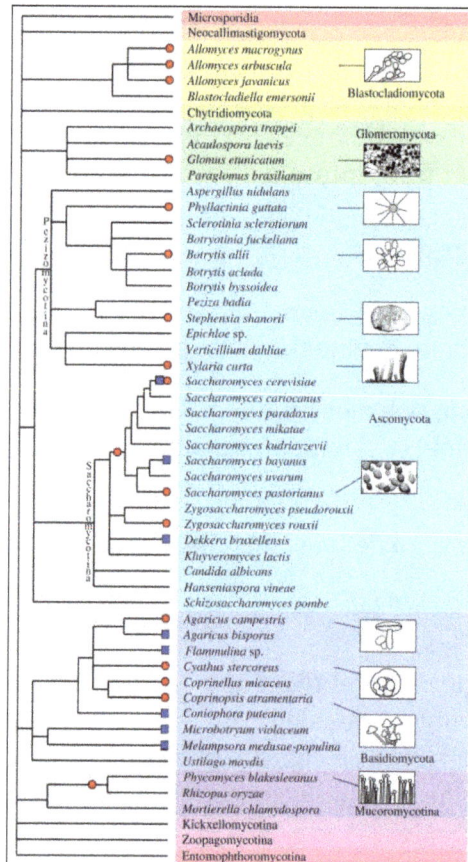

Schematiphylogeny of the fungi. Red circles indicate polyploidy, blue squares indicate hybridization. From Albertin and Marullo, 2012

Besides plants and animals, the evolutionary history of various fungal species is dotted by past and recent whole-genome duplication events. Several examples of polyploids are known:

- autopolyploid: the aquatifungi of genus *Allomyces*, some *Saccharomyces cerevisiae* strains used in bakery, etc.

- allopolyploid: the widespread *Cyathus stercoreus*, the allotetraploid lager yeast *Saccharomyces pastorianus*, the allotriploid wine spoilage yeast *Dekkera bruxellensis*, etc.

- paleopolyploid: the human pathogen *Rhizopus oryzae*, the *Saccharomyces* genus, etc.

In addition, polyploidy is frequently associated with hybridization and reticulate evolution that appear to be highly prevalent in several fungal taxa. Indeed, homoploid speciation (*i.e.*, hybrid speciation without a change in chromosome number) has been evidenced for some fungal species (*e.g.*, the basidiomycota *Microbotryum violaceum*).

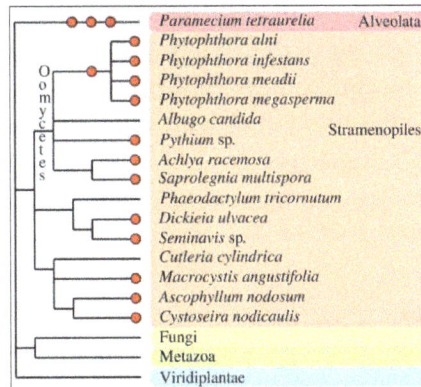

Schematiphylogeny of the Chromalveolata. Red circles indicate polyploidy, blue squares indicate hybridization. From Albertin and Marullo, 2012

As for plants and animals, fungal hybrids and polyploids display structural and functional modifications compared to their progenitors and diploid counterparts. In particular, the structural and functional outcomes of polyploid *Saccharomyces* genomes strikingly reflect the evolutionary fate of plant polyploid ones. Large chromosomal rearrangements leading to chimerichromosomes have been described, as well as more punctual genetimodifications such as gene loss. The homoealleles of the allotetraploid yeast *S. pastorianus* show unequal contribution to the transcriptome. Phenotypidiversification is also observed following polyploidization and/or hybridization in fungi, producing the fuel for natural selection and subsequent adaptation and speciation.

Chromalveolata

Other eukaryotitaxa have experienced one or more polyploidization events during their evolutionary history. The oomycetes, which are non-true fungi members, contain several examples of paleopolyploid and polyploid species, such as within the *Phytophthora* genus. Some species of brown algae (Fucales, Laminariales and diatoms) contain apparent polyploid genomes. In the Alveolata group, the remarkable species *Paramecium tetraurelia* underwent three successive rounds of whole-genome duplication and established itself as a major model for paleopolyploid studies.

Terminology

Autopolyploidy

Autopolyploids are polyploids with multiple chromosome sets derived from a single species. Autopolyploids can arise from a spontaneous, naturally occurring genome doubling, like the potato. Others might form following fusion of 2n gametes (unreduced gametes). Bananas and apples can be found as autotriploids. Autopolyploid plants typically display polysomiinheritance, and therefore have low fertility, but may be propagated clonally.

Allopolyploidy

Allopolyploids are polyploids with chromosomes derived from different species. Precisely it is the result of multiplying the chromosome number in an F1 hybrid. *Triticale* is an example of an allopolyploid, having six chromosome sets, allohexaploid, four from wheat (*Triticum turgidum*) and two from rye (*Secale cereale*). *Amphidiploids* are a type of allopolyploids (they are allotetraploid, containing the diploid chromosome sets of both parents). Some of the best examples of allopolyploids come from the Brassicas, and the Triangle of U describes the relationships between the three common diploid Brassicas (*B. oleracea, B. rapa,* and *B. nigra*) and three allotetraploids (*B. napus, B. juncea,* and *B. carinata*) derived from hybridization among the diploids.

Paleopolyploidy

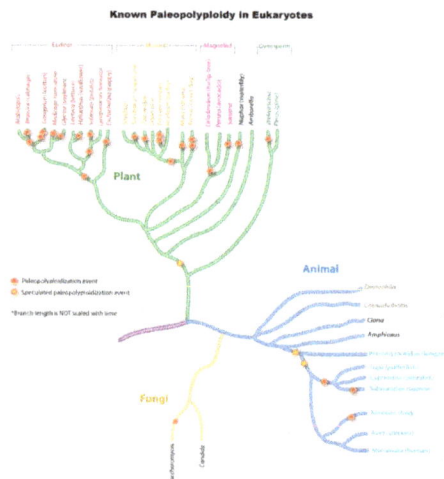

This phylogenetitree shows the relationship between the best-documented instances of paleopolyploidy in eukaryotes.

Ancient genome duplications probably occurred in the evolutionary history of all life. Duplication events that occurred long ago in the history of various evolutionary lineages can be difficult to detect because of subsequent diploidization (such that a polyploid starts to behave cytogenetically as a diploid over time) as mutations and gene translations gradually make one copy of each chromosome unlike the other copy. Over time, it is also common for duplicated copies of genes to accumulate mutations and become inactive pseudogenes.

In many cases, these events can be inferred only through comparing sequenced genomes. Examples of unexpected but recently confirmed ancient genome duplications include baker's yeast (*Saccharomyces cerevisiae*), mustard weed/thale cress (*Arabidopsis thaliana*), rice (*Oryza sativa*), and an early evolutionary ancestor of the vertebrates (which includes the human lineage) and another near the origin of the teleost fishes. Angiosperms (flowering plants) have paleopolyploidy in their ancestry. All eukaryotes probably have experienced a polyploidy event at some point in their evolutionary history.

Karyotype

A karyotype is the characteristichromosome complement of a eukaryote species. The preparation and study of karyotypes is part of cytology and, more specifically, cytogenetics.

Although the replication and transcription of DNA is highly standardized in eukaryotes, the same cannot be said for their karotypes, which are highly variable between species in chromosome number and in detailed organization despite being constructed out of the same macromolecules. In some cases, there is even significant variation within species. This variation provides the basis for a range of studies in what might be called evolutionary cytology.

Paralogous

The term is used to describe the relationship between duplicated genes or portions of chromosomes that derived from a common ancestral DNA. Paralogous segments of DNA may arise spontaneously by errors during DNA replication, copy and paste transposons, or whole genome duplications.

Homologous

The term is used to describe the relationship of similar chromosomes that pair at mitosis and meiosis. In a diploid, one homolog is derived from the male parent (sperm) and one is derived from the female parent (egg). During meiosis and gametogenesis, homologous chromosomes pair and exchange genetimaterial by recombination, leading to the production of sperm or eggs with chromosome haplotypes containing novel genetivariation.

Homoeologous

The term *homoeologous*, also spelled *homeologous*, is used to describe the relationship of similar chromosomes or parts of chromosomes brought together following inter-species hybridization and allopolyploidization, and whose relationship was completely homologous in an ancestral species. In allopolyploids, the homologous chromosomes within each parental sub-genome should pair faithfully during meiosis, leading to disomiinheritance; however in some allopolyploids, the homoeologous chromosomes of the parental genomes may be nearly as similar to one another as the homologous chromosomes, leading to tetrasomiinheritance (four chromosomes pairing at meiosis), intergenomirecombination, and reduced fertility.

Example of Homoeologous Chromosomes

Durum wheat is the result of the inter-species hybridization of two diploid grass species *Triticum urartu* and *Aegilops speltoides*. Both diploid ancestors had two sets of 7 chromosomes, which were similar in terms of size and genes contained on them. Durum wheat contains two sets of chromosomes derived from *Triticum urartu* and two sets of chromosomes derived from *Aegilops speltoides*. Each chromosome pair derived from the *Triticum urartu* parent is homoeologous to the opposite chromosome pair derived from the *Aegilops speltoides* parent, though each chromosome pair unto itself is homologous.

Polyploidization and Speciation

Polyploidization is a mechanism of sympatrispeciation because polyploids are usually unable to interbreed with their diploid ancestors. An example is the plant *Mimulus peregrinus*. Sequencing confirmed that this species originated from *M. x robertsii*, a sterile triploid hybrid between *M. guttatus* and *M. luteus,* both of which have been introduced and naturalised in the United Kingdom. New populations of *M. peregrinus* arose on the Scottish mainland and the Orkney Islands via genome duplication from local populations of *M. x robertsii.*

Sport (Botany)

Foliage of a dwarf Alberta spruce (*Picea glauca* var. *albertiana* 'Conica'), with a branch showing reversion to the normal Alberta white spruce growth habit of larger leaves and longer internodes.

In botany, a sport or bud sport, traditionally called ***lusus***, is a part of a plant (usually a woody plant, but sometimes a herb) that shows morphological differences from the rest of the plant. Sports may differ by foliage shape or color, flowers, or branch structure.

Sports with desirable characteristics are often propagated vegetatively to form new cultivars that retain the characteristics of the new morphology. Such selections are often prone to "reversion", meaning that part or all of the plant reverts to its original form. An example of a bud sport is the nectarine, at least some of which developed as a bud sport from peaches. Other common fruits resulting from a sport mutation are the red Anjou pear and the 'Pink Lemonade' lemon which is a sport of the "Eureka" lemon.

References

- Campbell, Neil A.; Brad Williamson; Robin J. Heyden (2006). Biology: Exploring Life. Boston, Massachusetts: Pearson Prentice Hall. ISBN 0-13-250882-6.

- Alberts, Bruce; Alexander Johnson; Julian Lewis; Martin Raff; Keith Roberts; Peter Walter (1994). Molecular Biology of the Cell. New York: Garland Publishing Inc. ISBN 0-8153-3218-1.

- Voet, Donald; Judith G. Voet; Charlotte W. Pratt (2006). Fundamentals of Biochemistry, 2nd Edition. John Wiley and Sons, Inc. pp. 547, 556. ISBN 0-471-21495-7.

- Margulis, Lynn; Sagan, Dorion (1986). Origins of Sex. Three Billion Years of Genetic Recombination. New Haven: Yale University Press. pp. 69–71, 87. ISBN 0 300 03340 0.

- Stryer, Lubert (1995). "Citric acid cycle.". In: Biochemistry. (Fourth ed.). New York: W.H. Freeman and Company. pp. 509–527, 569–579, 614–616, 638–641, 732–735, 739–748, 770–773. ISBN 0 7167 2009 4.

- Siegel GJ, Agranoff BW, Fisher SK, Albers RW, Uhler MD, eds. (1999). Basic Neurochemistry (6 ed.). Lippincott Williams & Wilkins. ISBN 0-397-51820-X. Illustrations by Lorie M. Gavulic CS1 maint: Uses editors parameter (link)

- Jones, Daniel (2003) [1917], Peter Roach, James Hartmann and Jane Setter, eds., English Pronouncing Dictionary, Cambridge: Cambridge University Press, ISBN 3-12-539683-2 CS1 maint: Uses editors parameter (link)

- Wise, edited by Robert R.; Hoober, J. Kenneth (2006). The structure and function of plastids. Dordrecht: Springer. pp. 3–21. ISBN 978-1-4020-4061-0.

- Sandelius, Anna Stina (2009). The Chloroplast: Interactions with the Environment. Springer. p. 18. ISBN 978-

3-540-68696-5.

- Burgess, Jeremy (1989). An introduction to plant cell development. Cambridge: Cambridge university press. p. 62. ISBN 0-521-31611-1.

- Burgess,, Jeremy (1989). An introduction to plant cell development (Pbk. ed.). Cambridge: Cambridge university press. p. 46. ISBN 0-521-31611-1.

- J D, Rochaix (1998). The molecular biology of chloroplasts and mitochondria in Chlamydomonas. Dordrecht [u.a.]: Kluwer Acad. Publ. pp. 550–565. ISBN 978-0-7923-5174-0.

- Steer, Brian E.S. Gunning, Martin W. (1996). Plant cell biology : structure and function. Boston, Mass.: Jones and Bartlett Publishers. p. 24. ISBN 0-86720-504-0.

- Roberts, editor, Keith (2007). Handbook of plant science. Chichester, West Sussex, England: Wiley. p. 16. ISBN 978-0-470-05723-0.

Genetic Engineering: An Integrated Study

The alteration caused in the genome of an organism by using biotechnology is known as genetic engineering. Genetic engineering techniques permit variation of the DNA of living organisms. There are many methods present in genetic engineering which have been listed in the chapter. It can best be comprehended in confluence with the major topics listed in the following section.

Genetic Engineering

Genetic engineering, also called genetic modification, is the direct manipulation of an organism's genome using biotechnology. It is a set of technologies used to change the genetic makeup of cells, including the transfer of genes within and across species boundaries to produce improved or novel organisms. New DNA may be inserted in the host genome by first isolating and copying the genetic material of interest using molecular cloning methods to generate a DNA sequence, or by synthesizing the DNA, and then inserting this construct into the host organism. Genes may be removed, or "knocked out", using a nuclease. Gene targeting is a different technique that uses homologous recombination to change an endogenous gene, and can be used to delete a gene, remove exons, add a gene, or introduce point mutations.

An organism that is generated through genetic engineering is considered to be a genetically modified organism (GMO). The first GMOs were bacteria generated in 1973 and GM mice in 1974. Insulin-producing bacteria were commercialized in 1982 and genetically modified food has been sold since 1994. GloFish, the first GMO designed as a pet, was first sold in the United States in December 2003.

Genetic engineering techniques have been applied in numerous fields including research, agriculture, industrial biotechnology, and medicine. Enzymes used in laundry detergent and medicines such as insulin and human growth hormone are now manufactured in GM cells, experimental GM cell lines and GM animals such as mice or zebrafish are being used for research purposes, and genetically modified crops have been commercialized.

IUPAC definition

Process of inserting new genetic information into existing cells in order to modify a specific organism for the purpose of changing its characteristics.

Definition

Genetic engineering alters the genetic make-up of an organism using techniques that remove heritable material or that introduce DNA prepared outside the organism either directly into the host or into a cell that is then fused or hybridized with the host. This involves using recombinant nucleic acid (DNA or RNA) techniques to form new combinations of heritable genetic material followed by the incorporation of that material either indirectly through a vector system or directly through micro-injection, macro-injection and micro-encapsulation techniques.

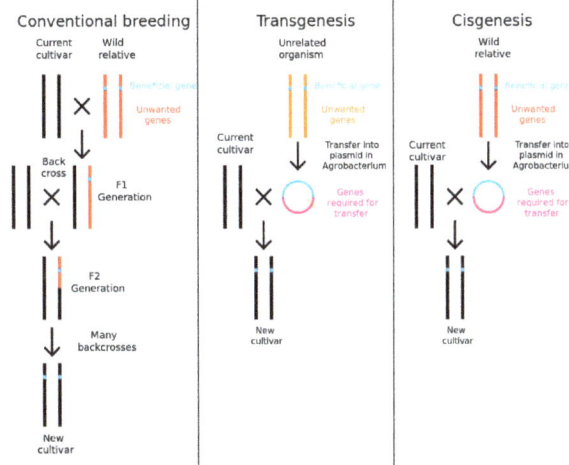

Comparison of conventional plant breeding with transgenic and cisgenic genetic modification.

Genetic engineering does not normally include traditional animal and plant breeding, in vitro fertilisation, induction of polyploidy, mutagenesis and cell fusion techniques that do not use recombinant nucleic acids or a genetically modified organism in the process. However the European Commission has also defined genetic engineering broadly as including selective breeding and other means of artificial selection. Cloning and stem cell research, although not considered genetic engineering, are closely related and genetic engineering can be used within them. Synthetic biology is an emerging discipline that takes genetic engineering a step further by introducing artificially synthesized material from raw materials into an organism.

If genetic material from another species is added to the host, the resulting organism is called transgenic. If genetic material from the same species or a species that can naturally breed with the host is used the resulting organism is called cisgenic. Genetic engineering can also be used to remove genetic material from the target organism, creating a gene knockout organism. In Europe genetic modification is synonymous with genetic engineering while within the United States of America it can also refer to conventional breeding methods. The Canadian regulatory system is based on whether a product has novel features regardless of method of origin. In other words, a product is regulated as genetically modified if it carries some trait not previously found in the species whether it was generated using traditional breeding methods (e.g., selective breeding, cell fusion, mutation breeding) or genetic engineering. Within the scientific community, the term *genetic engineering* is not commonly used; more specific terms such as *transgenic* are preferred.

Genetically Modified Organisms

Plants, animals or micro organisms that have changed through genetic engineering are termed genetically modified organisms or GMOs. Bacteria were the first organisms to be genetically modified. Plasmid DNA containing new genes can be inserted into the bacterial cell and the bacteria will then express those genes. These genes can code for medicines or enzymes that process food and other substrates. Plants have been modified for insect protection, herbicide resistance, virus resistance, enhanced nutrition, tolerance to environmental pressures and the production of edible vaccines. Most commercialised GMO's are insect resistant and/or herbicide tolerant crop plants. Genetically modified animals have been used for research, model animals and the production of agricultural or pharmaceutical products.

The genetically modified animals include animals with genes knocked out, increased susceptibility to disease, hormones for extra growth and the ability to express proteins in their milk.

History

Humans have altered the genomes of species for thousands of years through selective breeding, or artificial selection as contrasted with natural selection, and more recently through mutagenesis. Genetic engineering as the direct manipulation of DNA by humans outside breeding and mutations has only existed since the 1970s. The term "genetic engineering" was first coined by Jack Williamson in his science fiction novel *Dragon's Island*, published in 1951 – one year before DNA's role in heredity was confirmed by Alfred Hershey and Martha Chase, and two years before James Watson and Francis Crick showed that the DNA molecule has a double-helix structure – though the general concept of direct genetic manipulation was explored in rudimentary form in Stanley G. Weinbaum's 1936 science fiction story *Proteus Island*.

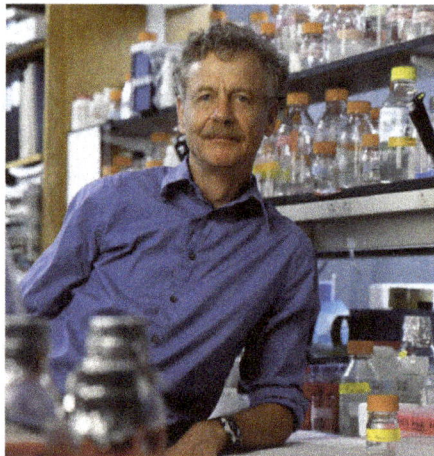

In 1974 Rudolf Jaenisch created the first GM animal.

In 1972, Paul Berg created the first recombinant DNA molecules by combining DNA from the monkey virus SV40 with that of the lambda virus. In 1973 Herbert Boyer and Stanley Cohen created the first transgenic organism by inserting antibiotic resistance genes into the plasmid of an *E. coli* bacterium. A year later Rudolf Jaenisch created a transgenic mouse by introducing foreign DNA into its embryo, making it the world's first transgenic animal. These achievements led to concerns in the scientific community about potential risks from genetic engineering, which were first discussed in depth at the Asilomar Conference in 1975. One of the main recommendations from this meeting was that government oversight of recombinant DNA research should be established until the technology was deemed safe.

In 1976 Genentech, the first genetic engineering company, was founded by Herbert Boyer and Robert Swanson and a year later the company produced a human protein (somatostatin) in *E.coli*.

Genentech announced the production of genetically engineered human insulin in 1978. In 1980, the U.S. Supreme Court in the *Diamond v. Chakrabarty* case ruled that genetically altered life could be patented. The insulin produced by bacteria, branded humulin, was approved for release by the Food and Drug Administration in 1982.

In the 1970s graduate student Steven Lindow of the University of Wisconsin–Madison with D.C. Arny and C. Upper found a bacterium he identified as *P. syringae* that played a role in ice nucleation, and in 1977 he discovered a mutant ice-minus strain. Dr. Lindow (who is now a plant pathologist at the University of California-Berkeley) later successfully created a recombinant ice-minus strain. In 1983, a biotech company, Advanced Genetic Sciences (AGS) applied for U.S. government authorization to perform field tests with the ice-minus strain of *P. syringae* to protect crops from frost, but environmental groups and protestors delayed the field tests for four years with legal challenges. In 1987, the ice-minus strain of *P. syringae* became the first genetically modified organism (GMO) to be released into the environment when a strawberry field and a potato field in California were sprayed with it. Both test fields were attacked by activist groups the night before the tests occurred: "The world's first trial site attracted the world's first field trasher".

The first field trials of genetically engineered plants occurred in France and the USA in 1986, tobacco plants were engineered to be resistant to herbicides. The People's Republic of China was the first country to commercialize transgenic plants, introducing a virus-resistant tobacco in 1992. In 1994 Calgene attained approval to commercially release the Flavr Savr tomato, a tomato engineered to have a longer shelf life. In 1994, the European Union approved tobacco engineered to be resistant to the herbicide bromoxynil, making it the first genetically engineered crop commercialized in Europe. In 1995, Bt Potato was approved safe by the Environmental Protection Agency, after having been approved by the FDA, making it the first pesticide producing crop to be approved in the USA. In 2009 11 transgenic crops were grown commercially in 25 countries, the largest of which by area grown were the USA, Brazil, Argentina, India, Canada, China, Paraguay and South Africa.

In 2010, scientists at the J. Craig Venter Institute created the first synthetic genome and inserted it into an empty bacterial cell. The resulting bacterium, named Synthia, could replicate and produce proteins. In 2014, a bacterium was developed that replicated a plasmid containing a unique base pair, creating the first organism engineered to use an expanded genetic alphabet.

Process

The first step is to choose and isolate the gene that will be inserted into the genetically modified organism. The gene can be isolated using restriction enzymes to cut DNA into fragments and gel electrophoresis to separate them out according to length. Polymerase chain reaction (PCR) can also be used to amplify up a gene segment, which can then be isolated through gel electrophoresis. If the chosen gene or the donor organism's genome has been well studied it may be present in a genetic library. If the DNA sequence is known, but no copies of the gene are available, it can be artificially synthesized.

The gene to be inserted into the genetically modified organism must be combined with other genetic elements in order for it to work properly. The gene can also be modified at this stage for better expression or effectiveness. As well as the gene to be inserted most constructs contain a promoter and terminator region as well as a selectable marker gene. The promoter region initiates transcription of the gene and can be used to control the location and level of gene expression, while the terminator region ends transcription. The selectable marker, which in most cases confers antibiotic resistance to the organism it is expressed in, is needed to determine which cells are transformed with the new gene. The constructs are made using recombinant DNA techniques, such as restriction digests, ligations and molecular cloning. The manipulation of the DNA generally occurs within a plasmid.

The most common form of genetic engineering involves inserting new genetic material randomly within the host genome. Other techniques allow new genetic material to be inserted at a specific location in the host genome or generate mutations at desired genomic loci capable of knocking out endogenous genes. The technique of gene targeting uses homologous recombination to target desired changes to a specific endogenous gene. This tends to occur at a relatively low frequency in plants and animals and generally requires the use of selectable markers. The frequency of gene targeting can be greatly enhanced with the use of engineered nucleases such as zinc finger nucleases, engineered homing endonucleases, or nucleases created from TAL effectors.

In addition to enhancing gene targeting, engineered nucleases can also be used to introduce mutations at endogenous genes that generate a gene knockout.

Transformation

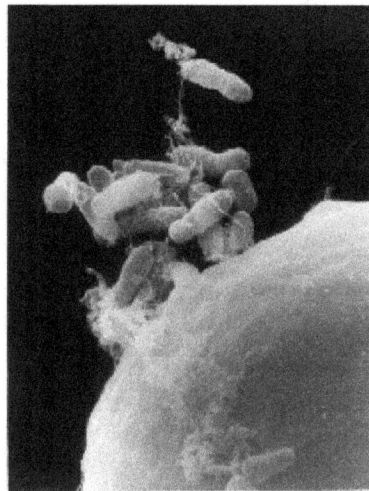

A. tumefaciens attaching itself to a carrot cell

Only about 1% of bacteria are naturally capable of taking up foreign DNA. However, this ability can be induced in other bacteria via stress (e.g. thermal or electric shock), thereby increasing the cell membrane's permeability to DNA; up-taken DNA can either integrate with the genome or exist as extrachromosomal DNA. DNA is generally inserted into animal cells using microinjection, where it can be injected through the cell's nuclear envelope directly into the nucleus or through the use of viral vectors. In plants the DNA is generally inserted using *Agrobacterium*-mediated recombination or biolistics.

In *Agrobacterium*-mediated recombination, the plasmid construct contains T-DNA, DNA which is responsible for insertion of the DNA into the host plants genome. This plasmid is transformed into *Agrobacterium* containing no plasmids prior to infecting the plant cells. The *Agrobacterium* will then naturally insert the genetic material into the plant cells. In biolistics transformation particles of gold or tungsten are coated with DNA and then shot into young plant cells or plant embryos. Some genetic material will enter the cells and transform them. This method can be used on plants that are not susceptible to *Agrobacterium* infection and also allows transformation of plant plastids. Another transformation method for plant and animal cells is electroporation. Electroporation involves subjecting the plant or animal cell to an electric shock, which can make the cell membrane permeable to plasmid DNA. In some cases the electroporated cells will incorporate the DNA into their genome. Due to the damage caused to the cells and DNA the transformation efficiency of biolistics and electroporation is lower than agrobacterial mediated transformation and microinjection.

As often only a single cell is transformed with genetic material the organism must be regenerated from that single cell. As bacteria consist of a single cell and reproduce clonally regeneration is not necessary. In plants this is accomplished through the use of tissue culture. Each plant species has different requirements for successful regeneration through tissue culture. If successful an adult plant is produced that contains the transgene in every cell. In animals it is necessary to ensure that the inserted DNA is present in the embryonic stem cells. Selectable markers are used to easily differentiate transformed from untransformed cells. These markers are usually present in the transgenic organism, although a number of strategies have been developed that can remove the selectable marker from the mature transgenic plant. When the offspring is produced they can be screened for the presence of the gene. All offspring from the first generation will be heterozygous for the inserted gene and must be mated together to produce a homozygous animal.

Further testing uses PCR, Southern hybridization, and DNA sequencing is conducted to confirm that an organism contains the new gene. These tests can also confirm the chromosomal location and copy number of the inserted gene. The presence of the gene does not guarantee it will be expressed at appropriate levels in the target tissue so methods that look for and measure the gene products (RNA and protein) are also used. These include northern hybridization, quantitative RT-PCR, Western blot, immunofluorescence, ELISA and phenotypic analysis. For stable transformation the gene should be passed to the offspring in a Mendelian inheritance pattern, so the organism's offspring are also studied.

Genome Editing

Genome editing is a type of genetic engineering in which DNA is inserted, replaced, or removed from a genome using artificially engineered nucleases, or "molecular scissors." The nucleases create specific double-stranded breaks (DSBs) at desired locations in the genome, and harness the cell's endogenous mechanisms to repair the induced break by natural processes of homologous recombination (HR) and nonhomologous end-joining (NHEJ). There are currently four families of engineered nucleases: meganucleases, zinc finger nucleases (ZFNs), transcription activator-like effector nucleases (TALENs), and the Cas9-guideRNA system (adapted from the CRISPR prokarotic immune system). In contrast to artificial genome editing natural genome editing occurs through viral and sub-viral agents competent in identification of genetic syntax structures for insertion/deletion processes with the result of conserved selection processes.

Applications

Genetic engineering has applications in medicine, research, industry and agriculture and can be used on a wide range of plants, animals and micro organisms.

Medicine

In medicine, genetic engineering has been used in manufacturing drugs, to create model animals and do laboratory research, and in gene therapy.

Manufacturing

Genetic engineering is used to mass-produce insulin, human growth hormones, follistim (for treating infertility), human albumin, monoclonal antibodies, antihemophilic factors, vaccines and many other drugs. Mouse hybridomas, cells fused together to create monoclonal antibodies, have been humanised through genetic engineering to create human monoclonal antibodies. Genetically engineered viruses are being developed that can still confer immunity, but lack the infectious sequences.

Research

Genetic engineering is used to create animal models of human diseases. Genetically modified mice are the most common genetically engineered animal model. They have been used to study and model cancer (the oncomouse), obesity, heart disease, diabetes, arthritis, substance abuse, anxiety, aging and Parkinson disease. Potential cures can be tested against these mouse models. Also genetically modified pigs have been bred with the aim of increasing the success of pig to human organ transplantation.

Gene Therapy

Gene therapy is the genetic engineering of humans, generally by replacing defective genes with effective ones. This can occur in somatic tissue or germline tissue.

Somatic gene therapy has been studied in clinical research in several diseases, including X-linked SCID, chronic lymphocytic leukemia (CLL), and Parkinson's disease. In 2012, Glybera became the first gene therapy treatment to be approved for clinical use in either Europe or the United States after its endorsement by the European Commission.

With regard to germline gene therapy, the scientific community has been opposed to attempts to alter genes in humans in inheritable ways using biotechnology since the technology was first introduced, and the caution has continued as the technology has progressed. With the advent of new techniques like CRISPR, in March 2015 scientists urged a worldwide ban on clinical use of gene editing technologies to edit the human genome in a way that can be inherited. In April 2015, Chinese researchers sparked controversy when they reported results of basic research experiments in which they edited the DNA of non-viable human embryos using CRISPR. In December 2015, scientists of major world academies called for a moratorium on inheritable human genome edits, including those related to CRISPR-Cas9 technologies.

There are also ethical concerns should the technology be used not just for treatment, but for enhancement, modification or alteration of a human beings' appearance, adaptability, intelligence, character or behavior. The distinction between cure and enhancement can also be difficult to establish. Transhumanists consider the enhancement of humans desirable.

Research

Knockout mice

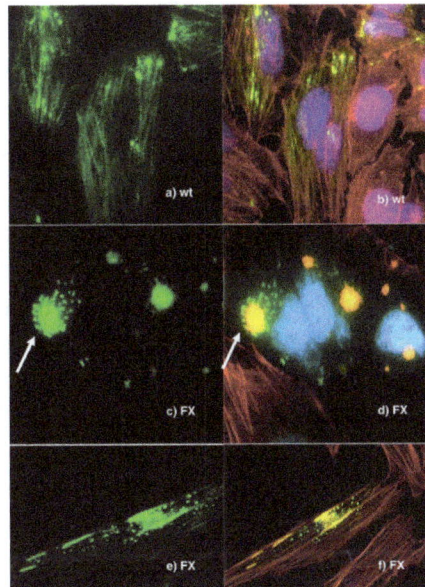

Human cells in which some proteins are fused with green fluorescent protein to allow them to be visualised

Genetic engineering is an important tool for natural scientists. Genes and other genetic information from a wide range of organisms are transformed into bacteria for storage and modification, creating genetically modified bacteria in the process. Bacteria are cheap, easy to grow, clonal, multiply quickly, relatively easy to transform and can be stored at -80 °C almost indefinitely. Once a gene is isolated it can be stored inside the bacteria providing an unlimited supply for research.

Organisms are genetically engineered to discover the functions of certain genes. This could be the effect on the phenotype of the organism, where the gene is expressed or what other genes it interacts with. These experiments generally involve loss of function, gain of function, tracking and expression.

- Loss of function experiments, such as in a gene knockout experiment, in which an organism is engineered to lack the activity of one or more genes. A knockout experiment involves the creation and manipulation of a DNA construct *in vitro*, which, in a simple knockout, consists of a copy of the desired gene, which has been altered such that it is non-functional. Embryonic stem cells incorporate the altered gene, which replaces the already present functional copy. These stem cells are injected into blastocysts, which are implanted into surrogate mothers. This allows the experimenter to analyze the defects caused by this mutation and thereby determine the role of particular genes. It is used especially frequently in developmental biology. Another method, useful in organisms such as Drosophila (fruit fly), is to induce mutations in a large population and then screen the progeny for the desired mutation. A similar process can be used in both plants and prokaryotes. Loss of function tells whether or not a protein is required for a function, but it does not always mean it's sufficient, especially if a function requires multiple proteins and lose the said function if one protein is missing.

- Gain of function experiments, the logical counterpart of knockouts. These are sometimes performed in conjunction with knockout experiments to more finely establish the function of the desired gene. The process is much the same as that in knockout engineering, except that the construct is designed to increase the function of the gene, usually by providing extra copies of the gene or inducing synthesis of the protein more frequently. Gain of function is used to tell whether or not a protein is sufficient for a function, but it does not always mean it's required. Especially when dealing with genetic/functional redundancy.

- Tracking experiments, which seek to gain information about the localization and interaction of the desired protein. One way to do this is to replace the wild-type gene with a 'fusion' gene, which is a juxtaposition of the wild-type gene with a reporting element such as green fluorescent protein (GFP) that will allow easy visualization of the products of the genetic modification. While this is a useful technique, the manipulation can destroy the function of the gene, creating secondary effects and possibly calling into question the results of the experiment. More sophisticated techniques are now in development that can track protein products without mitigating their function, such as the addition of small sequences that will serve as binding motifs to monoclonal antibodies.

- Expression studies aim to discover where and when specific proteins are produced. In these experiments, the DNA sequence before the DNA that codes for a protein, known as a gene's promoter, is reintroduced into an organism with the protein coding region replaced by a reporter gene such as GFP or an enzyme that catalyzes the production of a dye. Thus the time and place where a particular protein is produced can be observed. Expression studies can be taken a step further by altering the promoter to find which pieces are crucial for the proper expression of the gene and are actually bound by transcription factor proteins; this process is known as promoter bashing.

Industrial

Using genetic engineering techniques one can transform microorganisms such as bacteria or yeast, or transform cells from multicellular organisms such as insects or mammals, with a gene coding for a useful protein, such as an enzyme, so that the transformed organism will overexpress the desired protein. One can manufacture mass quantities of the protein by growing the transformed organism in bioreactor equipment using techniques of industrial fermentation, and then purifying the protein. Some genes do not work well in bacteria, so yeast, insect cells, or mammalians cells, each a eukaryote, can also be used. These techniques are used to produce medicines such as insulin, human growth hormone, and vaccines, supplements such as tryptophan, aid in the production of food (chymosin in cheese making) and fuels. Other applications involving genetically engineered bacteria being investigated involve making the bacteria perform tasks outside their natural cycle, such as making biofuels, cleaning up oil spills, carbon and other toxic waste and detecting arsenic in drinking water. Certain genetically modified microbes can also be used in biomining and bioremediation, due to their ability to extract heavy metals from their environment and incorporate them into compounds that are more easily recoverable.

Experimental, Lab Scale Industrial Applications

In materials science, a genetically modified virus has been used in an academic lab as a scaffold for assembling a more environmentally friendly lithium-ion battery.

Bacteria have been engineered to function as sensors by expressing a fluorescent protein under certain environmental conditions.

Agriculture

Bt-toxins present in peanut leaves (bottom image) protect it from extensive damage caused by European corn borer larvae (top image).

One of the best-known and controversial applications of genetic engineering is the creation and use of genetically modified crops or genetically modified organisms, such as genetically modified fish, which are used to produce genetically modified food and materials with diverse uses. There are four main goals in generating genetically modified crops.

One goal, and the first to be realized commercially, is to provide protection from environmental threats, such as cold (in the case of Ice-minus bacteria), or pathogens, such as insects or viruses, and/or resistance to herbicides. There are also fungal and virus resistant crops developed or in development. They have been developed to make the insect and weed management of crops easier and can indirectly increase crop yield.

Another goal in generating GMOs is to modify the quality of produce by, for instance, increasing the nutritional value or providing more industrially useful qualities or quantities. The Amflora potato, for example, produces a more industrially useful blend of starches. Cows have been engineered to produce more protein in their milk to facilitate cheese production. Soybeans and canola have been genetically modified to produce more healthy oils.

Another goal consists of driving the GMO to produce materials that it does not normally make. One example is "pharming", which uses crops as bioreactors to produce vaccines, drug intermediates, or drug themselves; the useful product is purified from the harvest and then used in the standard pharmaceutical production process. Cows and goats have been engineered to express drugs and other proteins in their milk, and in 2009 the FDA approved a drug produced in goat milk.

Another goal in generating GMOs, is to directly improve yield by accelerating growth, or making the organism more hardy (for plants, by improving salt, cold or drought tolerance). Salmon have been genetically modified with growth hormones to increase their size.

The genetic engineering of agricultural crops can increase the growth rates and resistance to different diseases caused by pathogens and parasites. This is beneficial as it can greatly increase the production of food sources with the usage of fewer resources that would be required to host the world's growing populations. These modified crops would also reduce the usage of chemicals, such as fertilizers and pesticides, and therefore decrease the severity and frequency of the damages produced by these chemical pollution.

Ethical and safety concerns have been raised around the use of genetically modified food. A major safety concern relates to the human health implications of eating genetically modified food, in particular whether toxic or allergic reactions could occur. Gene flow into related non-transgenic crops, off target effects on beneficial organisms and the impact on biodiversity are important environmental issues. Ethical concerns involve religious issues, corporate control of the food supply, intellectual property rights and the level of labeling needed on genetically modified products.

Conservation

Genetic engineering has potential applications in conservation and natural areas management. For example, gene transfer through viral vectors has been proposed as a means of controlling invasive species as well as vaccinating threatened fauna from disease. Transgenic trees have also been suggested as a way to confer resistance to pathogens in wild populations. With the increasing risks of maladaptation in organisms as a result of climate change and other perturbations, facilitated adaptation through gene tweaking could be one solution to reducing extinction risks. Applications of genetic engineering in conservation are thus far mostly theoretical and have yet to be put into practice. Further experimentation will be necessary to gauge the benefits and costs of such practices.

BioArt and Entertainment

Genetic engineering is also being used to create BioArt. Some bacteria have been genetically engineered to create black and white photographs.

Genetic engineering has also been used to create novelty items such as lavender-colored carnations, blue roses, and glowing fish.

Regulation

The regulation of genetic engineering concerns the approaches taken by governments to assess and manage the risks associated with the development and release of genetically modified crops. There are differences in the regulation of GM crops between countries, with some of the most marked differences occurring between the USA and Europe. Regulation varies in a given country depending on the intended use of the products of the genetic engineering. For example, a crop not intended for food use is generally not reviewed by authorities responsible for food safety. Starting in the late 1980s, guidance on assessing the safety of genetically engineered plants and food emerged from organizations including the FAO and WHO.

Controversy

Critics have objected to use of genetic engineering per se on several grounds, including ethical concerns, ecological concerns, and economic concerns raised by the fact GM techniques and GM organisms are subject to intellectual property law. GMOs also are involved in controversies over GM food with respect to whether food produced from GM crops is safe, whether it should be labeled, and whether GM crops are needed to address the world's food needs. These controversies have led to litigation, international trade disputes, and protests, and to restrictive regulation of commercial products in some countries.

Genetic Engineering Techniques

Genetic engineering techniques enable modification of the DNA of living organisms. A variety of editing techniques have been developed since DNA's structure was first discovered.

Targets

Bacteria are commonly engineered for research purposes. Typically this is through transformation to add a plasmid containing a gene of interest, but editing of the chromosome is also used. Plants and animals have both been genetically modified for research, agricultural and medical applications. In plants, the most widely inserted genes provide herbicide resistance or insecticidal properties. In animals, the most widely used are growth hormone genes. Finally, genetically modified viruses (such as retroviruses and lentiviruses) are also used as viral vectors to transfer target genes to another organism in gene therapy.

Procedure

The first step involves choosing and isolating the gene that will be inserted into/removed from the genetically modified organism.

The gene must generally be combined with a promoter and terminator region as well as a selectable marker gene.

Then the genes must be spliced into the target's DNA. For animals, the gene must be inserted into embryonic stem cells.

The resulting organism must have the presence of the target gene confirmed.

First generation offspring are heterozygous, requiring them to be inbred to create the homozygous pattern necessary for stable inheritance. Homozygosity must be confirmed in second generation specimens, which then become the final product.

History

Human directed genetic manipulation began with the domestication of plants and animals through artificial selection in about 12,000 BC. Various techniques were developed to aid in breeding and selection. Hybridization was one way rapid changes in an organisms makeup could be introduced. Hybridization most likely first occurred when humans first grew similar, yet slightly different plants in close proximity. Some plants were able to be propagated by vegetative cloning. X-rays were first used to deliberately mutate plants in 1927. Between 1927 and 2007, more than 2,540 genetically mutated plant varieties had been produced using x-rays.

It wasn't until the mid 1800s that DNA and genes were discovered, which would form the basis of modern genetic manipulation. Genetic inheritance was first discovered by Gregor Mendel in 1865 following experiments crossing peas. In 1928 Frederick Griffith proved the existence of a "transforming principle" involved in inheritance, which was identified as DNA in 19. Frederick Sanger developed a method for sequencing DNA in 1977, greatly increasing the genetic information available to researchers.

As well as discovering how DNA works, tools had to be developed that allowed it to be manipulated. In 1970 Hamilton Smiths lab discovered restriction enzymes, enabling scientists to isolate genes from an organism's genome. DNA ligases, that join broken DNA together, had been discovered earlier in 1967 and by combining the two enzymes it was possible to "cut and paste" DNA sequences to create recombinant DNA. Plasmids, discovered in 1952, became important tools for transferring information between cells and replicating DNA sequences. Polymerase chain reaction (PCR), developed by Kary Mullis in 1983, allowed small sections of DNA to be amplified and aided identification and isolation of genetic material.

As well as manipulating the DNA, techniques had to be developed for its insertion (known as transformation) into an organism's genome. Griffiths experiment had already shown that some bacteria had the ability to naturally uptake and express foreign DNA. Artificial competence was induced in *Escherichia coli* in 1970 by treating them with calcium chloride solution ($CaCl_2$). Transformation using electroporation was developed in the late 1980s, increasing the efficiency and bacterial range. In 1907 a bacterium that caused plant tumors, *Agrobacterium tumefaciens*, had been discovered and in the early 1970s it was found that the bacteria inserted its DNA into plants using a Ti plasmid. By removing the genes in the plasmid that caused the tumor and adding in novel genes researchers were able to infect plants with *A. tumefaciens* and let the bacteria insert their chosen DNA into the genomes of the plants.

An important part of genetic engineering is to identify useful genes to transform into the genetically modified organism. The bacteria *Bacillus thuringiensis* was first discovered in 1901 as the causative agent in the death of silkworms. Due to these insecticidal properties the bacteria was used as an biological insecticide, commercially developed in 1938. The cry proteins were discovered to provide the insecticidal activity in 1956 and by the 1980s scientists had successfully cloned the gene coding for this protein and expressed it in plants. The gene that provides resistance to the glyphosate herbicide was found, after seven years searching, in the outflow pipe of a Monsanto roundup manufacturing facility. In animals the majority of genes used are growth hormone genes.

Libraries

Target genes can be cloned from a DNA segment after the creation of a DNA library. The libraries generally cover the organism's genome multiple times and its size will depend on how large the genome is.

Techniques

Gene Isolation

The DNA is first digested with a random digestion method, commonly by cutting the DNA with restriction enzymes (enzymes that cut DNA). A partial restriction digest cuts only some of the restriction sites, resulting in overlapping DNA fragment lengths. The DNA fragments are put into individual plasmid vectors and grown inside bacteria. Once in the bacteria the plasmid is copied as the bacteria divides. To determine if a useful gene is present on a particular fragment the bacterial library is screened for the desired phenotype. If the phenotype is detected then it is possible that the bacteria contains the target gene. If the gene does not have a detectable phenotype or a DNA library does not contain the correct gene, other methods can be used to isolate it. If the position of the gene can be determined using molecular markers then chromosome walking is one way to isolate the correct DNA fragment. If the gene expresses close homology to a known gene in another species, then it could be isolated by searching for genes in the library that closely match the known gene.

If the DNA sequence of the gene and the organism is known, restriction enzymes can cut the DNA on either side of the gene and gel electrophoresis can sort the fragments according to length. The DNA band at the correct size should contain the gene, where it can be excised from the gel. Polymerase chain reaction (PCR) can be used to amplify the gene, which can then be isolated through gel electrophoresis. It is also possible to synthesize the gene.

Additional Material

The gene to be inserted into the genetically modified organism must be combined with other genetic elements in order for it to work properly. The gene can also be modified at this stage for better expression or effectiveness. As well as the gene to be inserted most constructs contain a promoter and terminator region as well as a selectable marker gene. The promoter region initiates transcription of the gene and can be used to control the location and level of gene expression, while the terminator region ends transcription. The selectable marker, which in most cases confers antibiotic resistance to the organism it is expressed in, is needed to determine which cells are transformed with the new gene. The constructs are made using recombinant DNA techniques, such as restriction digests, ligations and molecular cloning.

Gene Targeting

Gene targeting uses homologous recombination to target desired changes to a specific endogenous gene. This tends to occur at a relatively low frequency in plants and animals and generally requires the use of selectable markers. The success of gene targeting can be enhanced with the use of engineered nucleases such as zinc finger nucleases, engineered homing endonucleases, transcription activator-like effector nuclease. or CRISPR. Engineered nucleases can also introduce mutations at endogenous genes that generate a gene knockout.

Transformation

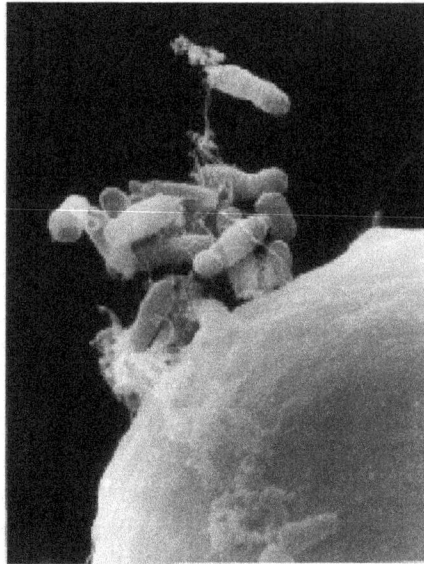

A. tumefaciens attaching itself to a carrot cell

About 1% of bacteria are naturally able to take up foreign DNA but it can be induced in other bacteria. Stressing the bacteria with a heat shock or an electric shock can make the cell membrane permeable to DNA that may then incorporate into the genome or exist as extrachromosomal DNA. DNA is generally inserted into animal cells using microinjection, where it can be injected through the cells nuclear envelope directly into the nucleus or through the use of viral vectors. In plants the DNA is generally inserted using *Agrobacterium*-mediated recombination or biolistics.

In *Agrobacterium*-mediated recombination the plasmid construct must also contain T-DNA. *Agrobacterium* naturally inserts DNA from a tumor inducing plasmid into any susceptible plant that it infects, causing crown gall disease. The T-DNA region of this plasmid is responsible for insertion of the DNA. The DNA to be inserted is cloned into a binary vector that contains T-DNA and can be grown in both *E. coli* and *Agrobacterium*. Once the binary vector is constructed the plasmid is transformed into *Agrobacterium* containing no plasmids and plant cells are infected. The *Agrobacterium* naturally inserts the genetic material into the plant cells.

In biolistic particles of gold or tungsten are coated with DNA and then shot into young plant cells or plant embryos. Some genetic material enters the cells and transforms them. This method can be used on plants that are not susceptible to *Agrobacterium* infection and also allows transformation of plant plastids.

Another transformation method for plant and animal cells is electroporation, which involves subjecting cells to an electric shock, which can make the cell membrane permeable to plasmid DNA. In some cases the electroporated cells will incorporate the DNA. Due to the associated cell and DNA damage, the transformation efficiency of biolistics and electroporation is lower than with agrobacteria and microinjection.

Selection

Not all the organism's cells will be transformed with the new genetic material; typically a selectable marker is used to differentiate transformed from untransformed cells. Cells that have been successfully transformed with the DNA it will also contain the marker gene. By growing the cells in the presence of an antibiotic or chemical that selects or marks the cells expressing that gene, it is possible to separate modified from unmodified cells. Another screening method involves a DNA probe that sticks only to the inserted gene. Multiple strategies can remove the marker from the mature plant.

Regeneration

As often only a single cell is transformed with genetic material the modified organism must be grown from that single cell. Bacteria consist of a single cell and reproduce clonally, so regeneration is not necessary for them. In plants this is accomplished through the use of tissue culture. Each plant species has different requirements for successful regeneration. If successful, the technique produces an adult plant that contains the transgene in every cell.

In animals it is necessary to ensure that the inserted DNA is present in embryonic stem cells. Offspring can be screened for the gene. All offspring from the first generation will be heterozygous for the inserted gene and must be inbred to produce a homozygous specimen.

Confirmation

The finding that a recombinant organism contains the inserted genes is not usually sufficient to ensure that they will be appropriately expressed in the intended tissues. To confirm the presence of the gene, PCR, Southern hybridization and DNA sequencing are employed to determine the chromosomal location and number of gene copies.

To assess gene expression, transcription, RNA processing patterns and expression and localization of protein product(s) must usually be assessed, using methods including northern hybridization, quantitative RT-PCR, Western blot, immunofluorescence and phenotypic analysis. When appropriate, the organism's offspring are studied to confirm that the trans-gene and associated phenotype are stably inherited.

In some cases further generations must be produced and confirmed, to ensure the absence of undesirable traits in the modified organism. For hybrid products such as maize, the modified organism is crossbred with other cultivars that possess required traits.

Gene Knockin

In molecular cloning and biology, a knock-in (or gene knock-in) refers to a genetic engineering method that involves the one-for-one substitution of DNA sequence information with a wild-type copy in a genetic locus or the insertion of sequence information not found within the locus. Typically, this is done in mice since the technology for this process is more refined and there is a high degree of shared sequence complexity between mice and humans. The difference between knock-in technology and traditional transgenic techniques is that a knock-in involves a gene inserted into a specific locus, and is thus a "targeted" insertion.

A common use of knock-in technology is for the creation of disease models. It is a technique by which scientific investigators may study the function of the regulatory machinery (e.g. promoters) that governs the expression of the natural gene being replaced. This is accomplished by observing the new phenotype of the organism in question. The BACs and YACs are used in this case so that large fragments can be transferred.

Technique

Gene knockin originated as a slight modification of the original knockout technique developed by Martin Evans, Oliver Smithies, and Mario Capecchi. Traditionally, knockin techniques have relied on homologous recombination to drive targeted gene replacement, although other methods using a transposon-mediated system to insert the target gene have been developed. The use of *loxP* flanking sites that become excised upon expression of Cre recombinase with gene vectors is an example of this. Embryonic stem cells with the modification of interest are then implanted into a viable blastocyst, which will grow into a mature chimeric mouse with some cells having the original blastocyst cell genetic information and other cells having the modifications introduced to the embryonic stem cells. Subsequent offspring of the chimeric mouse will then have the gene knockin.

Gene knockin has allowed, for the first time, hypothesis-driven studies on gene modifications and resultant phenotypes. Mutations in the human p53 gene, for example, can induced by exposure to benzo(a)pyrene and the mutated copy of the p53 gene can be inserted into mouse genomes. Lung tumors observed in the knockin mice offer support for the hypothesis of BaP's carcinogenicity. More recent developments in knockin technique have allowed for pigs to have a gene for green fluorescent protein inserted with a CRISPR/Cas9 system, which allows for much more accurate and successful gene insertions. The speed of CRISPR/Cas9-mediated gene knockin also allows for biallelic modifications to some genes to be generated and the phenotype in mice observed in a single generation, an unprecedented timeframe.

Versus Gene Knockout

Knockin technology is different from knockout technology in that knockout technology aims to either delete part of the DNA sequence or insert irrelevant DNA sequence information to disrupt the expression of a specific genetic locus. Gene knockin technology, on the other hand, alters the genetic locus of interest via a one-for-one substitution of DNA sequence information or by the addition of sequence information that is not found on said genetic locus. A gene knockin therefore can be seen as a gain of function mutation and a gene knockout a loss of function mutation, but a gene knockin may also involve the substitution of a functional gene locus for a mutant phenotype that results in some loss of function.

Potential Applications

Because of the success of gene knockin methods thus far, many clinical applications can be envisioned. Knockin of sections of the human immunoglobulin gene into mice has already been shown to allow them to produce humanized antibodies that are therapeutically useful. It should be possible to modify stem cells in humans to restore targeted gene function in certain tissues, for example possibly correcting the mutant gamma-chain gene of the IL-2 receptor in hematopoietic stem cells to restore lymphocyte development in people with X-linked severe combined immunodeficiency.

Limitations

While gene knockin technology has proven to be a powerful technique for the generation of models of human disease and insight into proteins *in vivo*, numerous limitations still exist. Many of these are shared with the limitations of knockout technology. First, combinations of knockin genes lead to growing complexity in the interactions that inserted genes and their products have with other sections of the genome and can therefore lead to more side effects and difficult-to-explain phenotypes. Also, only a few loci, such as the ROSA26 locus have been characterized well enough where they can be used for conditional gene knockins, making combinations of reporter and transgenes in the same locus problematic. The biggest disadvantage of using gene knockin for human disease model generation is that mouse physiology is not identical to that of humans and human orthologs of proteins expressed in mice will often not wholly reflect the role of a gene in human pathology. This can be seen in mice produced with the ΔF508 fibrosis mutation in the CFTR gene, which accounts for more than 70% of the mutations in this gene for the human population and leads to cystic fibrosis. While ΔF508 CF mice do exhibit the processing defects characteristic of the human mutation, they do not display the pulmonary pathophysiological changes seen in humans and carry virtually no lung phenotype. Such problems could be ameliorated by the use of a variety of animal models, and pig models (pig lungs share many biochemical and physiological similarities with human lungs) have been generated in an attempt to better explain the activity of the ΔF508 mutation.

Genome Editing

Genome editing, or genome editing with engineered nucleases (GEEN) is a type of genetic engineering in which DNA is inserted, deleted or replaced in the genome of a living organism using engineered nucleases, or "molecular scissors." These nucleases create site-specific double-strand breaks (DSBs) at desired locations in the genome. The induced double-strand breaks are repaired through nonhomologous end-joining (NHEJ) or homologous recombination (HR), resulting in targeted mutations.

There are currently four families of engineered nucleases being used: meganucleases, zinc finger nucleases (ZFNs), transcription activator-like effector-based nucleases (TALEN), and the CRISPR-Cas system.

Genome editing was selected by *Nature Methods* as the 2011 Method of the Year. The CRISPR-Cas system was selected by *Science* as 2015 Breakthrough of the Year.

Concept

A common approach in modern biological research is to modify the DNA sequence (genotype) of an organism (or a single cell) and observe the impact of this change on the organism (phenotype). This approach is called reverse genetics and its significance for modern biology lies in its relative simplicity. This method contrasts with that of forward genetics, where a new phenotype is first observed and then its genetic basis is studied. This course is more complex because phenotypic changes are often a result of multiple genetic interactions.

Among the key requirements of reverse genetic analysis is the ability to modify the DNA sequence of the target organism. This can be achieved by:

- site-directed mutagenesis employing either phage- or polymerase chain reaction (PCR)-mediated methods and oligonucleotides containing the desired mutation. These methods are most useful in organisms with straightforward methods for the introduction and selection of genes of interest, such as bacteria and yeast.

- recombination based methods that utilize the natural ability of cells to exchange DNA between its own genetic information and an exogenous DNA. These methods have been made possible in yeast and mice.

These approaches have several drawbacks:

- PCR- and phage-mediated approaches are less successful in more complex organisms such as mammals, where delivery becomes more difficult.

- They also require stringent selection steps and thus the addition of selection-specific sequences, along with those incorporated into the DNA.

- Recombination-based methods can be quite inefficient - e.g. in mouse embryonic stem cells treated with donor DNA, only 1 in a million DNA molecules was incorporated at the desired position.

The use of other mutagenic techniques (such as P-element transgenesis in *Drosophila*) also has limitations, the major one being the randomness of incorporation and the possibility of affecting other genes and expression patterns.

Hence, genome editing with engineered nucleases is a promising new approach. This rapidly evolving technology overcomes these shortcomings and uses relatively simple concepts.

Double Stranded Breaks and their Repair

Fundamental to the use of nucleases in genome editing is the concept of DNA double stranded break (DSB) repair mechanisms. Two of the known DSB repair pathways that are essentially functional in all organisms are the non-homologous end joining (NHEJ) and homology directed repair (HDR).

NHEJ uses a variety of enzymes to directly join the DNA ends in a double-strand break. In contrast, in HDR, a homologous sequence is utilized as a template for regeneration of missing DNA sequence at the break point. The natural properties of these pathways form the very basis of nucleases based genome editing.

NHEJ is error-prone, and has been shown to cause mutations at the repair site in approximately 50% of DSB in mycobacteria, and also its low fidelity has been linked to mutational accumulation in leukemias. Thus if one is able to create a DSB at a desired gene in multiple samples, it is very likely that mutations will be generated at that site in some of the treatments because of errors created by the NHEJ infidelity.

On the other hand, the dependency of HDR on a homologous sequence to repair DSBs can be exploited by inserting a desired sequence within a sequence that is homologous to the flanking sequences of a DSB which, when used as a template by HDR system, would lead to the creation of the desired change within the genomic region of interest.

Despite the distinct mechanisms, the concept of the HDR based gene editing is in a way similar to that of homologous recombination based gene targeting. However, the rate of recombination is increased by at least three orders of magnitude when DSBs are created and HDR is at work thus making the HDR based recombination much more efficient and eliminating the need for stringent positive and negative selection steps. So based on these principles if one is able to create a DSB at a specific location within the genome, then the cell's own repair systems will help in creating the desired mutations.

Site-specific Double Stranded Breaks

Creation of a DSB in DNA should not be a challenging task as the commonly used restriction enzymes are capable of doing so. However, if genomic DNA is treated with a particular restriction endonuclease many DSBs will be created. This is a result of the fact that most restriction enzymes recognize a few base pairs on the DNA as their target and very likely that particular base pair combination will be found in many locations across the genome. To overcome this challenge and create site-specific DSB, three distinct classes of nucleases have been discovered and bioengineered to date. These are the Zinc finger nucleases (ZFNs), transcription-activator like effector nucleases (TALEN) and meganucleases. Below is a brief overview and comparison of these enzymes and the concept behind their development.

Current Engineered Nucleases

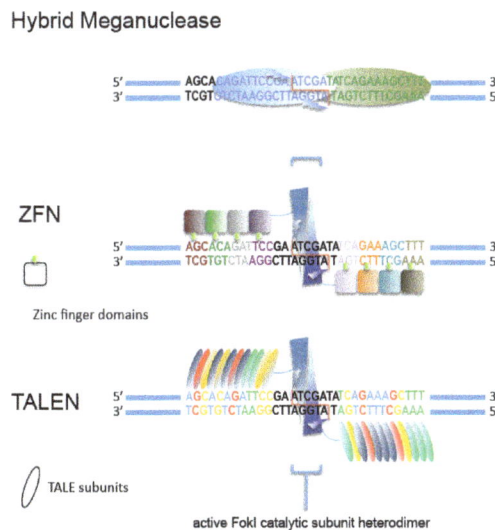

The current groups of engineered nucleases used for GEEN. Matching colors signify DNA recognition patterns

Meganucleases, found commonly in microbial species, have the unique property of having very long recognition sequences (>14bp) thus making them naturally very specific. This can be exploited to make site-specific DSB in genome editing; however, the challenge is that not enough meganucleases are known, or may ever be known, to cover all possible target sequences. To overcome this challenge, mutagenesis and high throughput screening methods have been used to create meganuclease variants that recognize unique sequences. Others have been able to fuse various meganucleases and create hybrid enzymes that recognize a new sequence. Yet others have attempted to alter the DNA interacting aminoacids of the meganuclease to design sequence specific meganucelases in a method named rationally designed meganuclease (US Patent 8,021,867 B2).

Meganucleases have the benefit of causing less toxicity in cells than methods such as ZFNs, likely because of more stringent DNA sequence recognition; however, the construction of sequence-specific enzymes for all possible sequences is costly and time consuming, as one is not benefiting from combinatorial possibilities that methods such as ZFNs and TALEN-based fusions utilize.

As opposed to meganucleases, the concept behind ZFNs and TALEN technology is based on a non-specific DNA cutting enzyme, which can then be linked to specific DNA sequence recognizing peptides such as zinc fingers and transcription activator-like effectors (TALEs). The key to this was to find an endonuclease whose DNA recognition site and cleaving site were separate from each other, a situation that is not common among restriction enzymes. Once this enzyme was found, its cleaving portion could be separated which would be very non-specific as it would have no recognition ability. This portion could then be linked to sequence recognizing peptides that could lead to very high specificity. A restriction enzyme with such properties is FokI. Additionally FokI has the advantage of requiring dimerization to have nuclease activity and this means the specificity increases dramatically as each nuclease partner would recognize a unique DNA sequence. To enhance this effect, FokI nucleases have been engineered that can only function as heterodimers and have increased catalytic activity. The heterodimer functioning nucleases would avoid the possibility of unwanted homodimer activity and thus increase specificity of the DSB. Although the nuclease portions of both ZFNs and TALEN constructs have similar properties, the difference between these engineered nucleases is in their DNA recognition peptide. ZFNs rely on Cys2-His2 zinc fingers and TALEN constructs on TALEs. Both of these DNA recognizing peptide domains have the characteristic that they are naturally found in combinations in their proteins. Cys2-His2 Zinc fingers typically happen in repeats that are 3 bp apart and are found in diverse combinations in a variety of nucleic acid interacting proteins such as transcription factors. TALEs on the other hand are found in repeats with a one-to-one recognition ratio between the amino acids and the recognized nucleotide pairs. Because both zinc fingers and TALEs happen in repeated patterns, different combinations can be tried to create a wide variety of sequence specificities. Zinc fingers have been more established in these terms and approaches such as modular assembly (where Zinc fingers correlated with a triplet sequence are attached in a row to cover the required sequence), OPEN (low-stringency selection of peptide domains vs. triplet nucleotides followed by high-stringency selections of peptide combination vs. the final target in bacterial systems), and bacterial one-hybrid screening of zinc finger libraries among other methods have been used to make site specific nucleases.

One recent improvement combines the DNA binding specificity of TALEs with the nuclease specificity of meganucleases; these "megaTALs" are compatible with all current technologies and may represent improvements on existing methods.

Precision of Engineered Nucleases

Because off-target activity of an active nuclease would have potentially dangerous consequences at the genetic and organismal levels, the precision of meganucleases, ZFNs, CRISPR, and TALEN-based fusions has been an active area of research. While variable figures have been reported, ZFNs tend to have more cytotoxicity than TALEN methods or RNA-guided nucleases, while TALEN and RNA-guided approaches tend to have the greatest efficiency and fewer off-target effects. Based on the maximum theoretical distance between DNA binding and nuclease activity, TALEN approaches result in the greatest precision.

Applications

Organisms	Genes	Methods of ZFN development
Gene disruption		
Fruitflies	yellow, rosy, brown	Modular assembly
Zebrafish	kdr	Bacteria one-hybrid
	golden, no tail	Two-finger modules
	trf2, dat, telomerase	OPEN
Human T cells	CCR5	Two-finger modules
Rats	Rab38, IgM, Il2rg	Two-finger modules
Gene correction		
Tobacco	SuRA, SuRB	OPEN
Arabidopsis thaliana	ABI4, KU80	Modular assembly
	TT4, ADH1	OPEN
Fruitflies	yellow, rosy, coilin, pask	Modular assembly
Human T cells	IL2RG	Two-finger modules
Gene addition		
Tobacco	Chitinase	Two-finger modules
Zea mays	Ipkl, Zein protein 15	Two-finger modules
Human ES cells	IL2RG, CCR5	Two-finger modules
	PIGA	OPEN

Over the past decade, efficient genome editing has been developed for a wide range of experimental systems ranging from plants to animals, often beyond clinical interest, and the method holds a promising future in becoming a standard experimental strategy in research labs. The recent generation of rat, zebrafish, maize and tobacco ZFN-mediated mutants and the improvements in TALEN-based approaches testify to the significance of the methods, and the list is expanding rapidly. Genome editing with engineered nucleases will likely contribute to many fields of life sciences from studying gene functions in plants and animals to gene therapy in humans. For instance, the field of synthetic biology which aims to engineer cells and organisms to perform novel functions, is likely to benefit from the ability of engineered nuclease to add or remove genomic elements and therefore create complex systems. In addition, gene functions can be studied using stem cells with engineered nucleases.

Listed below are some specific tasks this method can carry out:

- Targeted gene mutation

- Creating chromosome rearrangement

- Study gene function with stem cells

- Transgenic animals

- Endogenous gene labeling

- Targeted transgene addition

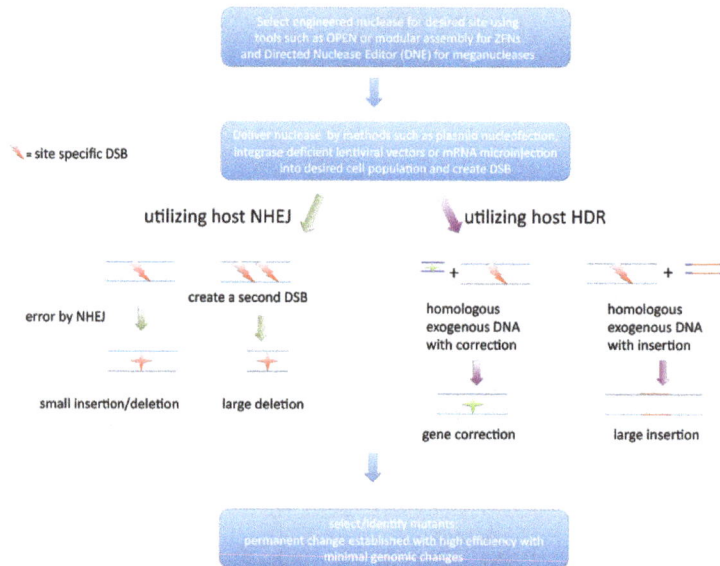

Overview of GEEN workflow and editing possibilities

Targeted Gene Modification in Plants

Genome editing using meganucleases, ZFNs, and TALEN provides a new strategy for genetic manipulation in plants and are likely to assist in the engineering of desired plant traits by modifying endogenous genes. For instance, site-specific gene addition in major crop species can be used for 'trait stacking' whereby several desired traits are physically linked to ensure their co-segregation during the breeding processes. Progress in such cases have been recently reported in *Arabidopsis thaliana* and *Zea mays*. In *Arabidopsis thaliana*, using ZFN-assisted gene targeting, two herbicide-resistant genes (tobacco acetolactate synthase SuRA and SuRB) were introduced to SuR loci with as high as 2% transformed cells with mutations. In Zea mays, disruption of the target locus was achieved by ZFN-induced DSBs and the resulting NHEJ. ZFN was also used to drive herbicide-tolerance gene expression cassette (PAT) into the targeted endogenous locus IPK1 in this case. Such genome modification observed in the regenerated plants has been shown to be inheritable and was transmitted to the next generation.

In addition, TALEN-based genome engineering has been extensively tested and optimized for use in plants. TALEN fusions have also been used to improve the quality of soybean oil products and to increase the storage potential of potatoes

Several optimizations need to be made in order to improve editing plant genomes using ZFN-mediated targeting. These include the reliable design and subsequent test of the nucleases, the absence of toxicity of the nucleases, the appropriate choice of the plant tissue for targeting, the routes of introduction or induction of enzyme activity, the lack of off-target mutagenesis, and a reliable detection of mutated cases.

Gene Therapy

The ideal gene therapy practice is that which replaces the defective gene with a normal allele at its natural location. This is advantageous over a virally delivered gene as there is no need to include the full coding sequences and regulatory sequences when only a small proportions of the gene needs to be altered as is often the case. The expression of the partially replaced genes is also more consistent with normal cell biology than full genes that are carried by viral vectors.

Gene targeting through ZFNs or TALEN-based approaches can also be used to modify defective genes at their endogenous chromosomal locations. Examples include the treatment of X-linked severe combined immunodeficiency (X-SCID) by ex vivo gene correction with DNA carrying the interleukin-2 receptor common gamma chain (IL-2Rγ) and the correction of Xeroderma pigmentosum mutations in vitro using TALEN. Insertional mutagenesis by the retroviral vector genome induced leukemia in some patients, a problem that is predicted to be avoided by these technologies. However, ZFNs may also cause off-target mutations, in a different way from viral transductions. Currently many measures are taken to improve off-target detection and ensure safety before treatment.

Recently, Sangamo BioSciences (SGMO) introduced the Delta 32 mutation (a suppressor of CCR5 gene which is a co-receptor for HIV-1 entry into T cells therefore enabling HIV infection) using Zinc Finger Nuclease (ZFN). Their results were presented at the 51st Interscience Conference on Antimicrobial Agents and Chemotherapy (ICAAC) held in Chicago from September 17–20, 2011. Researchers at SGMO mutated CCR5 in CD4+ T cells and subsequently produced an HIV-resistant T-cell population.

Gene editing is used to generate modified custom immune cells. For example, recent report indicated that T cells could be modified to inactivate the glucocorticoid receptor; the resulting immune cells are fully functional but resistant to the effects of commonly used corticosteroids. Similarly, scientists at Cellectis recently generated custom T-cells expressing chimeric antigen receptors using TALEN technology. These T-cells can be engineered to be resistant to anti-cancer drugs and to invoke immune responses against targets of interest.

The first clinical use of TALEN-based genome editing was in the treatment of CD19+ acute lymphoblastic leukemia in an 11-month old child. Modified donor T cells were engineered to attack the leukemia cells, to be resistant to Alemtuzumab, and to evade detection by the host immune system after introduction. A few weeks after therapy, the patient's condition improved. Though physicians are cautious, the patient has been in remission for several months following treatment.

Prospects and Limitations

In the future, an important goal of research into genome editing with engineered nucleases must be the improvement of the safety and specificity of the nucleases. For example, improving the ability to detect off-target events can improve our ability to learn about ways of preventing them. In addition, zinc-fingers used in ZFNs are seldom completely specific, and some may cause a toxic reaction. However, the toxicity has been reported to be reduced by modifications done on the cleavage domain of the ZFN.

In addition, research by Dana Carroll into modifying the genome with engineered nucleases has shown the need for better understanding of the basic recombination and repair machinery of DNA.

In the future, a possible method to identify secondary targets would be to capture broken ends from cells expressing the ZFNs and to sequence the flanking DNA using high-throughput sequencing.

Genome editing occurs also as a natural process without artificial genetic engineering. The agents that are competent to edit genetic codes are viruses or subviral RNA-agents.

Lastly, although GEEN has higher efficiency than many other methods in reverse genetics, it is still not highly efficient; in many cases less than half of the treated populations obtain the desired changes. For example, when one is planning to use the cell's NHEJ to create a mutation, the cell's HDR systems will also be at work correcting the DSB with lower mutational rates.

References

- McHugen, Alan (14 September 2000). "Chapter 1: Hors-d'oeuvres and entrees/What is genetic modification? What are GMOs?". Pandora's Picnic Basket. Oxford University Press. ISBN 978-0198506744.

- Panesar, Pamit et al (2010) "Enzymes in Food Processing: Fundamentals and Potential Applications", Chapter 10, I K International Publishing House, ISBN 978-9380026336

- Shiv Kant Prasad, Ajay Dash (2008). Modern Concepts in Nanotechnology, Volume 5. Discovery Publishing House. ISBN 9788183562966. CS1 maint: Uses authors parameter (link)

- Head, Graham; Hull, Roger H; Tzotzos, George T. (2009). Genetically Modified Plants: Assessing Safety and Managing Risk. London: Academic Pr. p. 244. ISBN 0-12-374106-8.

- Avise, John C. (2004). The hope, hype & reality of genetic engineering: remarkable stories from agriculture, industry, medicine, and the environment. Oxford University Press US. p. 22. ISBN 978-0-19-516950-8.

- Reece, Jane B.; Urry, Lisa A.; Cain, Michael L.; Wasserman, Steven A.; Minorsky, Peter V.; Jackson, Robert B. (2011). Campbell Biology Ninth Edition. San Francisco: Pearson Benjamin Cummings. p. 421. ISBN 0-321-55823-5.

- Corinne A. Michels (2002). "7". Genetic Techniques for Biological Research: A Case Study Approach. John Wiley & Sons. pp. 85–88. ISBN 0-471-89919-4.

- Graham Head; Hull, Roger H; Tzotzos, George T. (2009). Genetically Modified Plants: Assessing Safety and Managing Risk. London: Academic Pr. p. 244. ISBN 0-12-374106-8.

- Gibson, Greg (2009). A Primer Of Genome Science 3rd ed. Sunderland, Massachusetts: Sinauer. pp. 301–302. ISBN 978-0-87893-236-8.

- Pollack, Andrew (19 November 2015). "Genetically Engineered Salmon Approved for Consumption". The New York Times. Retrieved 21 April 2016.

- Thomas; et al. (25 September 2013). "Ecology: Gene tweaking for conservation". Nature. 501: 485–6. doi:10.1038/501485a. PMID 24073449. Retrieved 16 May 2016.

Threat to Plant Genetics

Genetic erosion is a process where a threatened species of plant becomes extinct without a chance to breed with the same species. The term is used to describe the loss of specific genes or for even an entire species. Habitat fragmentation is another reason for genetic erosion. This chapter provides the reader with an in-depth understanding on genetic erosion.

Genetic Erosion

Genetic erosion is a process whereby an already limited gene pool of an endangered species of plant or animal diminishes even more when individuals from the surviving population die off without getting a chance to meet and breed with others in their endangered low population. The term is sometimes used in a narrow sense, such as when describing the loss of particular alleles or genes, as well as being used more broadly, as when referring to the loss of varieties or even whole species.

Genetic erosion occurs because each individual organism has many unique genes which get lost when it dies without getting a chance to breed. Low genetic diversity in a population of wild animals and plants leads to a further diminishing gene pool – inbreeding and a weakening immune system can then "fast track" that species towards eventual extinction.

All the endangered species of the world are plagued to varying degrees by genetic erosion, and most need a human-assisted breeding program to keep their population viable, thereby avoiding extinction over long time frames. The smaller the population is on a relative scale, the more magnified the effect of genetic erosion becomes, as weakened individuals from the few surviving members of the species are lost without getting a chance to breed.

Genetic erosion also gets compounded and accelerated by habitat fragmentation – today most endangered species live in smaller and smaller chunks of (fragmented) habitat, interspersed with human settlements and farmland, making it much more difficult to naturally meet and breed with others of their kind, so many die off without getting a chance to breed at all, and thus are unable to pass on their unique genes to the living population.

The gene pool of a species or a population is the complete set of unique alleles that would be found by inspecting the genetic material of every living member of that species or population. A large gene pool indicates extensive genetic diversity, which is associated with robust populations that can survive bouts of intense selection. Meanwhile, low genetic diversity can cause reduced biological fitness and increase the chance of extinction of that species or population.

Processes and Consequences

Population bottlenecks create shrinking gene pools, which leave fewer and fewer fertile mating partners. The genetic implications can be illustrated by considering the analogy of a high-stakes poker game with a crooked dealer. Consider that the game begins with a 52-card deck (representing high genetic diversity). Reduction of the number of breeding pairs with unique genes resembles the situation where the dealer deals only the same five cards over and over, producing only a few limited "hands".

As specimens begin to inbreed, both physical and reproductive congenital effects and defects appear more often. Abnormal sperm increases, infertility rises, and birthrates decline. "Most perilous are the effects on the immune defense systems, which become weakened and less and less able to fight off an increasing number of bacterial, viral, fungal, parasitic, and other disease-producing threats. Thus, even if an endangered species in a bottleneck can withstand whatever human development may be eating away at its habitat, it still faces the threat of an epidemic that could be fatal to the entire population."

Loss of Agricultural and Livestock Biodiversity

Genetic erosion in agricultural and livestock is the loss of biological genetic diversity – including the loss of individual genes, and the loss of particular recombinants of genes (or gene complexes) – such as those manifested in locally adapted landraces of domesticated animals or plants that have become adapted to the natural environment in which they originated.

The major driving forces behind genetic erosion in crops are variety replacement, land clearing, overexploitation of species, population pressure, environmental degradation, overgrazing, governmental policy, and changing agricultural systems. The main factor, however, is the replacement of local varieties of domestic plants and animals by other varieties or species that are non-local. A large number of varieties can also often be dramatically reduced when commercial varieties are introduced into traditional farming systems. Many researchers believe that the main problem related to agro-ecosystem management is the general tendency towards genetic and ecological uniformity imposed by the development of modern agriculture.

Prevention by Human Intervention, Modern Science and Safeguards

In Situ Conservation

With advances in modern bioscience, several techniques and safeguards have emerged to check the relentless advance of genetic erosion and the resulting acceleration of endangered species towards eventual extinction. However, many of these techniques and safeguards are too expensive yet to be practical, and so the best way to protect species is to protect their habitat and to let them live in it as naturally as possible.

Wildlife sanctuaries and national parks have been created to preserve entire ecosystems with all the web of species native to the area. Wildlife corridors are created to join fragmented habitats to enable endangered species to travel, meet, and breed with others of their kind. Scientific conservation and modern wildlife management techniques, with the expertise of scientifically trained staff, help manage these protected ecosystems and the wildlife found in them. Wild animals are also translocated and reintroduced to other locations physically when fragmented wildlife habitats are too far and isolated to be able to link together via a wildlife corridor, or when local extinctions have already occurred.

Ex Situ Conservation

Modern policies of zoo associations and zoos around the world have begun putting dramatically increased emphasis on keeping and breeding wild-sourced species and subspecies of animals in their registered endangered species breeding programs. These specimens are intended to have a chance to be reintroduced and survive back in the wild. The main objectives of zoos today have changed, and greater resources are being invested in breeding species and subspecies for then ultimate purpose of assisting conservation efforts in the wild. Zoos do this by maintaining extremely detailed scientific breeding records (*i.e.* studbooks)) and by loaning their wild animals to other zoos around the country (and often globally) for breeding, to safeguard against inbreeding by attempting to maximize genetic diversity however possible.

Costly (and sometimes controversial) ultra-modern ex-situ conservation techniques have emerged that aim to increase the genetic biodiversity on our planet, as well as the diversity in local gene pools. by guarding against genetic erosion. Modern concepts like seedbanks, sperm banks, and tissue banks have become much more commonplace and valuable. Sperm, eggs, and embryos can now be frozen and kept in banks, which are sometimes called "Modern Noah's Arks" or "Frozen Zoos". Cryopreservation techniques are used to freeze these living materials and keep them alive in perpetuity by storing them submerged in liquid nitrogen tanks at very low temperatures. Thus, preserved materials can then be used for artificial insemination, *in vitro* fertilization, embryo transfer, and cloning methodologies to protect diversity in the gene pool of critically endangered species.

It is today possible to save an endangered species from extinction by preserving only *parts* of specimens, such as tissues, sperm, eggs, etc. – even after the death of a critically endangered animal, or collected from one found freshly dead, in captivity or from the wild. A new specimen can then be "resurrected" with the help of cloning, so as to give it another chance to breed its genes into the living population of the respective threatened species. Resurrection of dead critically endangered wildlife specimens with the help of cloning is still being perfected, and is still too expensive to be practical, but with time and further advancements in science and methodology it may well become a routine procedure not to far into the future.

Recently, strategies for finding an integrated approach to *in situ* and *ex situ* conservation techniques have been given considerable attention, and progress is being made.

References

- G. Tyler Miller; Scott Spoolman (24 September 2008). Living in the Environment: Principles, Connections, and Solutions. Cengage Learning. pp. 279–. ISBN 978-0-495-55671-8. Retrieved 7 September 2010.

- Richardson, Edited by David M. (2000). Ecology and biogeography of Pinus. Cambridge, U.K. p. 371. ISBN 978-0-521-78910-3.

- Haspel, Tamar (9 May 2014). "Monocrops: They're a problem, but farmers aren't the ones who can solve it.". New York Times. Retrieved 12 January 2016.

- Dobzhansky, T; Pavlovsky, O (1957). "An experimental study of interaction between genetic drift and natural selection". Evolution. 11 (3): 311–319. doi:10.2307/2405795. Retrieved 21 April 2016.

- Williams, J.L. (2015-10-22). "The Value of Genome Mapping for the Genetic Conservation of Cattle". The Food and Agriculture Organization of the United Nations. Rome. Retrieved 2015-10-22.

- Burke MK, Dunham JP, et al. (Sep 2015). "Genome-wide analysis of a long-term evolution experiment with Drosophila". Nature. 467: 587–90. doi:10.1038/nature09352. PMID 20844486.

- "Using experimental evolution and next-generation sequencing to teach bench and bioinformatic skills". PeerJ PrePrints (3): e1674. 2015. doi:10.7287/peerj.preprints.1356v1.

- Schlötterer C, Tobler R, Kofler R, Nolte V (Nov 2014). "Sequencing pools of individuals - mining genome-wide polymorphism data without big funding". Nat. Rev. Genet. 15: 749–63. doi:10.1038/nrg3803. PMID 25246196.

- Hyman, Paul (2014). "Bacteriophage as instructional organisms in introductory biology labs". Bacteriophage. 4 (2): e27336. doi:10.4161/bact.27336. ISSN 2159-7081.

- "A Novel Laboratory Activity for Teaching about the Evolution of Multicellularity". The American Biology Teacher. 76 (2): 81–87. 2014. doi:10.1525/abt.2014.76.2.3. ISSN 0002-7685.

- Early Canid Domestication: The Fox Farm Experiment, p.2, by Lyudmila N. Trut, Ph.D., Retrieved February 19, 2011.

Evolution of Plant Genetics

Planting has been practiced by thousands of years. It can easily be traced to the beginning of human civilization. Modern scientific advances have enabled the study of plant genetics. This chapter helps the reader in understanding the evolution and history of plant breeding.

Experimental Evolution

Experimental evolution is the use of experiments or controlled field manipulations to explore evolutionary dynamics. Evolution may be observed in the laboratory as populations adapt to new environmental conditions and/or change by such stochastic processes as random genetic drift. With modern molecular tools, it is possible to pinpoint the mutations that selection acts upon, what brought about the adaptations, and to find out how exactly these mutations work. Because of the large number of generations required for adaptation to occur, evolution experiments are typically carried out with microorganisms such as bacteria, yeast or viruses, or other organisms with rapid generation times. However, laboratory studies with foxes and with rodents have shown that notable adaptations can occur within as few as 10-20 generations and experiments with wild guppies have observed adaptations within comparable numbers of generations. More recently, using experimental evolution followed by whole genome pooled sequencing, an approach known as Evolve and Resequence (E&R) is becoming popular in fruit flies.

History

Domestication and Breeding

This Chihuahua mix and Great Dane show the wide range of dog breed sizes created using artificial selection.

Unwittingly, humans have carried out evolution experiments for as long as they have been domesticating plants and animals. Selective breeding of plants and animals has led to varieties that differ dramatically from their original wild-type ancestors. Examples are the cabbage varieties, maize, or the large number of different dog breeds. The power of human breeding to create varieties with extreme differences from a single species was already recognized by Charles Darwin. In fact, he started out his book *The Origin of Species* with a chapter on variation in domestic animals. In this chapter, Darwin discussed in particular the pigeon.

Altogether at least a score of pigeons might be chosen, which if shown to an ornithologist, and he were told that they were wild birds, would certainly, I think, be ranked by him as well-defined species. Moreover, I do not believe that any ornithologist would place the English carrier, the short-faced tumbler, the runt, the barb, pouter, and fantail in the same genus; more especially as in each of these breeds several truly-inherited sub-breeds, or species as he might have called them, could be shown him. (...) I am fully convinced that the common opinion of naturalists is correct, namely, that all have descended from the rock-pigeon (*Columba livia*), including under this term several geographical races or sub-species, which differ from each other in the most trifling respects.

—Charles Darwin, The Origin of Species

Early Experimental Evolution

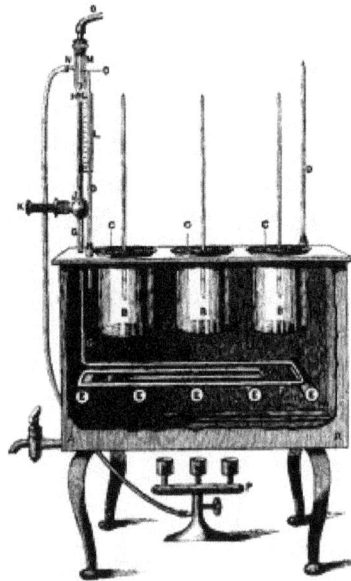

Drawing of the incubator used by Dallinger in his evolution experiments.

One of the first to carry out a controlled evolution experiment was William Dallinger. In the late 19th century, he cultivated small unicellular organisms in a custom-built incubator over a time period of seven years (1880–1886). Dallinger slowly increased the temperature of the incubator from an initial 60 °F up to 158 °F. The early cultures had shown clear signs of distress at a temperature of 73 °F, and were certainly not capable of surviving at 158 °F. The organisms Dallinger had in his incubator at the end of the experiment, on the other hand, were perfectly fine at 158 °F. However, these organisms would no longer grow at the initial 60 °F. Dallinger concluded that he had found evidence for Darwinian adaptation in his incubator, and that the organisms had adapted to live in a high-temperature environment. Unfortunately, Dallinger's incubator was accidentally destroyed in 1886, and Dallinger could not continue this line of research.

From the 1880s to 1980, experimental evolution was intermittently practiced by a variety of evolutionary biologists, including the highly influential Theodosius Dobzhansky. Like other experimental research in evolutionary biology during this period, much of this work lacked extensive replication and was carried out only for relatively short periods of evolutionary time.

Modern Experimental Evolution

Experimental evolution has been used in various formats to understand underlying evolutionary processes in a controlled system. Experimental evolution has been performed on multicellular and unicellular eukaryotes, prokaryotes, viruses. Similar works have also been performed by directed evolution of individual enzyme, ribozyme and replicator genes.

Fruit Flies

One of the first of a new wave of experiments using this strategy was the laboratory "evolutionary radiation" of *Drosophila melanogaster* populations that Michael R. Rose started in February, 1980. This system started with ten populations, five cultured at later ages, and five cultured at early ages. Since then more than 200 different populations have been created in this laboratory radiation, with selection targeting multiple characters. Some of these highly differentiated populations have also been selected "backward" or "in reverse," by returning experimental populations to their ancestral culture regime. Hundreds of people have worked with these populations over the better part of three decades. Much of this work is summarized in the papers collected in the book *Methuselah Flies*, listed below.

The early experiments in flies were limited to studying phenotypes but the molecular mechanisms, i.e., changes in DNA that facilitated such changes, could not be identified. This changed with genomics technology. Subsequently, Thomas Turner coined the term Evolve and Resequence (E&R) and several studies used E&R approach with mixed success (reviewed in and). One of the more interesting experimental evolution studies was conduced by Gabriel Haddad's group at UC San Diego, where Haddad and colleagues evolved flies to adapt to low oxygen environments, also known as hypoxia. After 200 generations, they used E&R approach to identify genomic regions that were selected by natural selection in the hypoxia adapted flies. More recent experiments are have started following up E&R predictions with RNAseq and genetic crosses. Such efforts in combining E&R with experimental validations should be powerful in identifying genes that regulate adaptation in flies.

Bacteria

Bacteria have short generation times, easily-sequenced genomes, and well understood biology. They are therefore commonly used for experimental evolution studies.

Lenski's E. Coli Experiment

One of the most widely known examples of laboratory bacterial evolution is the long-term *E.coli* experiment of Richard Lenski. On February 15, 1988, Lenski started growing lineages of *E. coli* in different growth conditions. When one of the flasks suddenly developed the ability to metabolize citrate from the growth medium aerobically and showed greatly increased growth, this provided a dramatic observation of evolution in action. The experiment continues to this day, and is now the longest-running controlled evolution experiment ever undertaken. Since the inception of the experiment, the bacteria have grown for more than 60,000 generations. Lenski and colleagues regularly publish updates on the status of the experiments.

Hyperswarming in P. Aeruginosa

In 2013, Joao Xavier published on an experiment in which *Pseudomonas aeruginosa*, when subjected to repeated rounds of conditions in which it needed to swarm to acquire food, developed the ability to "hyperswarm" at speeds 25% faster than baseline organisms, by developing multiple flagella, whereas the baseline organism has a single flagellum. Especially notable was that this development was astonishingly repeatable.

Laboratory House Mice

Mouse from the Garland selection experiment with attached running wheel and its rotation counter.

In 1998, Theodore Garland, Jr. and colleagues started a long-term experiment that involves selective breeding for high voluntary activity levels on running wheels. This experiment also continues to this day (> 65 generations). Mice from the four replicate "High Runner" lines evolved to run almost three times as many running-wheel revolutions per day compared with the four unselected control lines of mice, mainly by running faster than the control mice rather than running for more minutes/day.

Female mouse with her litter, from the Garland selection experiment.

The HR mice exhibit an elevated maximal aerobic capacity when tested on a motorized treadmill. They also exhibit alterations in motivation and the reward system of the brain. Pharmacological studies point to alterations in dopamine function and the endocannabinoid system. The High Runner lines have been proposed as a model to study human attention-deficit hyperactivity disorder (ADHD), and administration of Ritalin reduces their wheel running approximately to the levels of Control mice. Click here for a mouse wheel running video.

Other Examples

Stickleback fish have both marine and freshwater species, the freshwater species evolving since the last ice age. Fresh water species can survive colder temperatures. Scientists tested to see if they could reproduce this evolution of cold-tolerance by keeping marine sticklebacks in cold freshwater. It took the marine sticklebacks only three generations to evolve to match the 2.5 degree Celsius improvement in cold-tolerance found in wild freshwater sticklebacks.

Experimental Evolution for Teaching

Because of their rapid generation times microbes offer an opportunity to study microevolution in the classroom. A number of exercises involving bacteria and yeast teach concepts ranging from the evolution of resistance to the evolution of multicellularity. With the advent of next-generation sequencing technology it has become possible for students to conduct an evolutionary experiment, sequence the evolved genomes, and to analyze and interpret the results.

History of Plant Breeding

Plant breeding started with sedentary agriculture, particularly the domestication of the first agricultural plants, a practice which is estimated to date back 9,000 to 11,000 years. Initially, early human farmers selected food plants with particular desirable characteristics and used these as a seed source for subsequent generations, resulting in an accumulation of characteristics over time. In time however, experiments began with deliberate hybridization, the science and understanding of which was greatly enhanced by the work of Gregor Mendel. Mendel's work ultimately led to the new science of genetics. Modern plant breeding is applied genetics, but its scientific basis is broader, covering molecular biology, cytology, systematics, physiology, pathology, entomology, chemistry, and statistics (biometrics). It has also developed its own technology. Plant breeding efforts are divided into a number of different historical landmarks.

Early Plant Breeding

Domestication

Domestication of plants is an artificial selection process conducted by humans to produce plants that have more desirable traits than wild plants, and which renders them dependent on artificial usually enhanced environments for their continued existence. The practice is estimated to date back 9,000-11,000 years. Many crops in present-day cultivation are the result of domestication in ancient times, about 5,000 years ago in the Old World and 3,000 years ago in the New World. In the Neolithic period, domestication took a minimum of 1,000 years and a maximum of 7,000 years. Today, all principal food crops come from domesticated varieties. Almost all the domesticated plants used today for food and agriculture were domesticated in the centers of origin. In these centers there is still a great diversity of closely related wild plants, so-called crop wild relatives, that can also be used for improving modern cultivars by plant breeding.

Centers of origin of selected crops

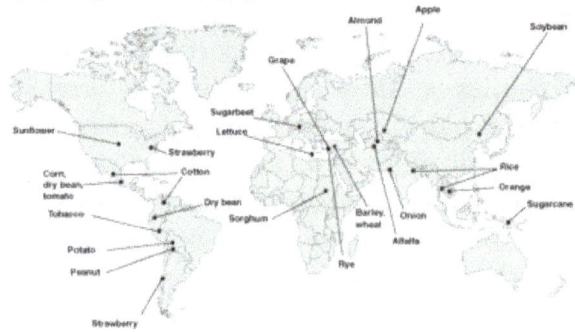

This map shows the sites of domestication for a number of crops. Places where crops were initially domesticated are called centers of origin

A plant whose origin or selection is due primarily to intentional human activity is called a cultigen, and a cultivated crop species that has evolved from wild populations due to selective pressures from traditional farmers is called a landrace. Landraces, which can be the result of natural forces or domestication, are plants or animals that are ideally suited to a particular region or environment. An example are the landraces of rice, *Oryza sativa* subspecies *indica*, which was developed in South Asia, and *Oryza sativa* subspecies *japonica*, which was developed in China.

For more on the mechanisms of domestication.

Columbian Exchange

Humans have traded useful plants from distant lands for centuries, and plant hunters have been sent to bring plants back for cultivation. Human agriculture has had two important results: the plants most favoured by humans came to be grown in many places and (2) gardens and farms have provided some opportunities for plants to interbreed that would not have been possible for their wild ancestors. Columbus's arrival in America in 1492 triggered unprecedented transfer of plant resources between Europe and the New World.

Scientific Plant Breeding

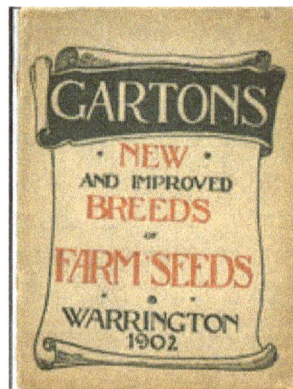

Garton's catalogue from 1902

Gregor Mendel's experiments with plant hybridization led to his laws of inheritance. This work became well known in the 1900s and formed the basis of the new science of genetics, which stimulated research by many plant scientists dedicated to improving crop production through plant breeding.

However, successful commercial plant breeding concerns began to be founded from the late 19th century. Gartons Agricultural Plant Breeders in England was established in the 1890s by John Garton, who was one of the first to cross-pollinate agricultural plants and commercialize the newly created varieties. He began experimenting with the artificial cross pollination firstly of cereal plants, then herbage species and root crops and developed far reaching techniques in plant breeding.

From 1904 to World War II in Italy, Nazareno Strampelli created a number of wheat hybrids. His work allowed Italy to increase crop production during the so-called "Battle for Grain" (1925–1940) and some varieties were exported to foreign countries, such as Argentina, Mexico, and China. Strampelli's work laid the foundations for Norman Borlaug and the Green Revolution.

Green Revolution

In 1908, George Harrison Shull described heterosis, also known as hybrid vigor. Heterosis describes the tendency of the progeny of a specific cross to outperform both parents. The detection of the usefulness of heterosis for plant breeding has led to the development of inbred lines that reveal a heterotic yield advantage when they are crossed. Maize was the first species where heterosis was widely used to produce hybrids.

By the 1920s, statistical methods were developed to analyze gene action and distinguish heritable variation from variation caused by environment. In 1933 another important breeding technique, cytoplasmic male sterility (CMS), developed in maize, was described by Marcus Morton Rhoades. CMS is a maternally inherited trait that makes the plant produce sterile pollen. This enables the production of hybrids without the need for labor-intensive detasseling.

These early breeding techniques resulted in large yield increase in the United States in the early 20th century. Similar yield increases were not produced elsewhere until after World War II, the Green Revolution increased crop production in the developing world in the 1960s. This remarkable improvement was based on three essential crops. First came the development of hybrid maize, then high-yielding and input-responsive "semi-dwarf wheat" (for which the CIMMYT breeder N.E. Borlaug received the Nobel prize for peace in 1970), and third came high-yielding "short statured rice" cultivars. Similarly notable improvements were achieved in other crops like sorghum and alfalfa.

Molecular Genetics and Bio-revolution

Intensive research in molecular genetics has led to the development of recombinant DNA technology (popularly called genetic engineering). Advancement in biotechnological techniques has opened many possibilities for breeding crops. Thus, while mendelian genetics allowed plant breeders to perform genetic transformations in a few crops, molecular genetics has provided the key to both the manipulation of the internal genetic structure, and the "crafting" of new cultivars according to a pre-determined plan.

Permissions

Index

www.ingramcontent.com/pod-product-compliance
Lightning Source LLC
Chambersburg PA
CBHW061253190326
41458CB00011B/3663